Advances in Stability of Metallic Implants

Advances in Stability of Metallic Implants

Editors

Changjiang Pan
Jingan Li

MDPI

Basel • Beijing • Wuhan • Barcelona • Belgrade • Novi Sad • Cluj • Manchester

Editors

Changjiang Pan
Faculty of Mechanical and Material Engineering
Huaiyin Institute of Technology
Huai'an
China

Jingan Li
School of Materials Science and Engineering
Zhengzhou University
Zhengzhou
China

Editorial Office
MDPI
St. Alban-Anlage 66
4052 Basel, Switzerland

This is a reprint of articles from the Special Issue published online in the open access journal *Metals* (ISSN 2075-4701) (available at: www.mdpi.com/journal/metals/special_issues/stability_implants).

For citation purposes, cite each article independently as indicated on the article page online and as indicated below:

Lastname, A.A.; Lastname, B.B. Article Title. *Journal Name* **Year**, *Volume Number*, Page Range.

ISBN 978-3-7258-0032-2 (Hbk)
ISBN 978-3-7258-0031-5 (PDF)
doi.org/10.3390/books978-3-7258-0031-5

Contents

metals

Editorial

Advances in Stability of Metallic Implants

Changjiang Pan [1,*] and Jingan Li [2]

1 Faculty of Mechanical and Material Engineering, Jiangsu Provincial Engineering Research Center for Biomaterials and Advanced Medical Devices, Huaiyin Institute of Technology, Huai'an 223003, China
2 School of Materials Science and Engineering, Henan Key Laboratory of Advanced Magnesium Alloys, Zhengzhou University, Zhengzhou 450001, China; lijingan@zzu.edu.cn
* Correspondence: panchangjiang@hyit.edu.cn

1. Introduction and Scope

Metallic implants have attracted extensive attention because of their importance in enhancing the quality of human lives and treating human diseases. There is a tremendous demand for metallic implants when organ defects or functional damage caused by trauma and disease occur. However, their instability in vivo caused by corrosion, limited mechanical properties, and poor biocompatibility remains a great challenge to be solved, because the stability of metallic implants has a close association with the clinical outcome of medical implants. Therefore, this topic has attracted increasing attention from researchers in the last several decades and great advances have been achieved.

The purpose of this Special Issue is to provide a platform to exhibit the researchers' latest findings, understandings, and insights into Advances in Stability of Metallic Implants. It is anticipated that this Special Issue can enhance the development of metallic implants that can fulfill the clinical requirements. Consequently, full-length articles, short communications and reviews regarding metallic implants and their biomaterials are covered in this special topic.

2. Contributions

In this Special Issue, we collected ten articles, including eight research papers and two review papers, which covered several aspects concerning Advances in Stability of Metallic Implants.

The paper by Alferi [1] investigated the influences of different metal alloys and their manufacturing technologies on anti-corrosion in SGF (standardized simulated gastric fluid). It was discovered that both raw material and the production parameters can influence the anti-corrosion properties, and the released nickel in SGF was positively related to the corrosion degree. On the other hand, the author found that not only the corrosion susceptibility characterized by gastric acid is inadequate and the low pH valve of the environment around the implant should be considered, but also the manufacturing process can abate the influence of the base material on corrosion susceptibility.

It is well known that the mechanical properties of medical implants have a significant impact on the in vivo performances after implantation, and biomimetic simulation and finite element analysis are effective methods for investigating the mechanical properties and in vivo stability of implants. Considering the structural features of natural human cervical, Cheng et al. [2] designed a new fixture for measuring the fatigue properties of artificial cervical disc (ACD) prostheses under different mechanical conditions, including flexion, extension, and lateral bending. The numerical simulations and mechanical experiments were carried out to characterize the equivalence between the designed fixture and the natural cervical sections, and the results indicated that this novel biomimetic fixture could reflect the biomechanical characters of the natural human cervical vertebrae with acceptable accuracy. In another work by Cheng et al. [3], a C2–C7 3D finite element model of the

Citation: Pan, C.; Li, J. Advances in Stability of Metallic Implants. *Metals* **2023**, *13*, 1718. https://doi.org/10.3390/met13101718

Received: 15 September 2023
Accepted: 27 September 2023
Published: 9 October 2023

cervical spine was constructed based on the data from cervical CT images. Different pure moment loads were used to simulate flexion/extension and the relative motion range between vertebral bodies. The stress of some segments under axial load was also investigated. The results were in accordance with the in vitro experimental data, indicating the validity of the new model.

In the work by Adelmann et al. [4], the authors applied laser powder bed fusion to prepare titanium alloys with different lattice structures, and the hot isostatic pressing was used as post-treatment. The properties of the as-prepared materials with different lattice structures were comparatively studied. It was discovered that hot isostatic pressing had no obvious impact on the strength when high loads and low cycle numbers were applied; however, all samples survived 10^6 cycles at low loads. As a result, after hot isostatic pressing, dode-thick and rhombic dodecahedrons with 2 mm and 1.5 mm lattice sizes were suitable for medical implants due to the high elasticity, high fracture stress, and high resistance against dynamic loads, which can meet requirements of the implants.

Magnesium (Mg) and its alloys have been extensively explored for implants such as cardiovascular stents owing to its biodegradability, but the in vivo fast degradation and poor biocompatibility hindered the applications in the clinic. Surface modification and further biofunctionalization are the popular ways to simultaneously enhance corrosion resistance and biocompatibility. The review by Wu et al. [5] first summarized various surface design strategies to control the corrosion rate of magnesium alloys. This work widely discussed the bioinspired surface designs to enhance anti-corrosion and realize biofunctionalization. The author thought that future work on the anti-corrosion of magnesium-based materials could benefit greatly from the bioinspired surfaces. Subsequently, to investigate the ways to improve the anti-corrosion properties and biocompatibility of magnesium alloys for cardiovascular stents, Meng et al. [6] applied four chemical treatments, including NaOH, HF, phosphoric acid (H_3PO_4) and phytic acid ($C_6H_{18}O_{24}P_6$) treatment, to produce the different chemical conversion coatings on AZ31 surface, and the influences of different layers on the corrosion properties and biocompatibility were investigated. It was discovered that all chemical treatments can improve the anti-corrosion properties to different degrees and significantly affect the surface wettability, which can cause different plasma protein adsorption behaviors. Furthermore, they found that all chemical treatments can enhance the hemocompatibility to varying degrees and the NaOH-treated surface had the best cytocompatibility to endothelial cells. Consequently, the NaOH treatment can be chosen as the pretreatment method for the biofunctionalization of the magnesium-based alloys to further improve the biocompatibility when applied as cardiovascular implants. In another paper regarding surface modification of magnesium alloy by Han et al. [7], ferulic acid (FA) was immobilized on the polydopamine (PDA) modified Mg-Zn-Y-Nd alloy to construct a PDA/FA coating to simultaneously enhance the corrosion resistance and biocompatibility. The outcomes suggested that the PDA/FA coating can be used to modify the medical Mg alloy to improve its performance. In the work by Sheng et al. [8], three Schiff bases were synthesized to construct the Schiff base coatings on the Mg-Zn-Y-Nd (ZE21B) surface by electrostatic spraying. The results indicated that both the single coating and the compound coating can reduce the corrosion rate of ZE21B alloy, and the compound coating had a synergistic effect in inhibiting corrosion, thus displaying the best anti-corrosion character.

Hu et al. [9] prepared Cu/Ti coatings containing different copper concentrations on titanium surface by the physical vapor deposition (PVD), and the effect of deposition bias on the properties of these coatings was further investigated. It was discovered that the deposition dispersion and copper concentration can be controlled by the number of copper sheets, but the bias voltage had no impact on the copper concentration of the Cu/Ti coating. As compared to control samples, Cu/Ti coating-modified samples showed better hemocompatibility. The best adsorption of bovine serum albumin (BSA) can be found on the coating when the bias voltage of -40 V is applied. The increase in copper flakes can reduce the adsorption of fibrinogen on the Cu/Ti-coated surface. Therefore, Cu/Ti coatings have the potential to enhance hemocompatibility and surface performances of cardiovascular biomaterials or implants.

As an important organic synthetic reagent in organic chemistry, the Schiff base has many excellent bioactivities, such as anti-tumor, anti-virus, antifungal, and antibacterial activities. Therefore, the coatings made of the Schiff base can improve the bioactivities of metal biomaterials. In the review paper by Zhang et al. [10], preparation, properties of the Schiff base, and advantages of Schiff base coatings were summarized. Through reviewing the advanced works concerning the functional coatings prepared by Schiff base and Schiff base reaction, and the extensive applications of Schiff base coatings in many aspects such as inhibition of corrosion, antibacterial, flame retardant, etc., the authors obtained four conclusions: (1) Schiff bases are extensively studied because of the synthetic flexibility, selectivity, structural similarity with natural bio-compounds, and their (−N-CH-) groups. (2) Schiff bases can be used to modify biomedical metals to create functional coatings to enhance biocompatibility, mechanical properties, and anti-corrosion properties. (3) An amino acid–paeonol Schiff base can offer the Mg alloy surface a self-healing property and enhance cytocompatibility to endothelial cells. (4) The application mechanisms of many Schiff bases and their compound coatings still need to be explored.

3. Conclusions and Outlook

Although many advances have been achieved in the stability of metallic implants in the past several decades, it is a problem far from being solved satisfactorily in clinical applications. This Special Issue makes some contributions to enhancing the comprehension of the stability of metallic implants and providing some methods to improve the mechanical properties and biocompatibility of the metallic implants. The guest editors hope that the data, the observations, and the methodology presented in these papers can greatly promote the development of metallic implants.

Funding: The research was funded by the National Natural Science Foundation of China (31870952), the Natural Science Foundation of the Jiangsu Higher Education Institutions of China (No. 21KJB430013), the Key Program for Natural Science Foundation of Jiangsu Higher Education Institutes of China (23KJA430004) and the Wenzhou Municipal Science and Technology Commission (No. ZY2020018).

Acknowledgments: As Guest Editors, we would like to thank all the authors' valuable contributions, and the time and hard work of the reviewers, editors, and the editorial staff of *Metals*.

Conflicts of Interest: The authors declare no conflict of interest.

References

1. Alferi, D.; Fojt, J.; Kristianova, E.; Edwards, D.W.; Laasch, H.U. Influence of the manufacturing process on the corrosion and mechanical behavior of Esophageal stents. *Metals* **2023**, *13*, 1542. [CrossRef]
2. Cheng, X.; Bai, J.; Wang, T. Biomimetic design of fatigue-testing fixture for artificial cervical disc prostheses. *Metals* **2023**, *13*, 299. [CrossRef]
3. Cheng, X.; Wang, T.; Pan, C. Finite element analysis and validation of segments C2-C7 of the cervical spine. *Metals* **2022**, *12*, 2056. [CrossRef]
4. Adelmann, B.; Hellmann, R. Mechanical properties of LPBF-built titanium lattice structures-A comparative study of as-built and hot isostatic pressed structures for medical implants. *Metals* **2022**, *12*, 2072. [CrossRef]
5. Wu, F.; Liu, Y.; Xu, J.; Pan, C. Bioinspired surface design for magnesium alloy with corrosion resistance. *Metals* **2022**, *12*, 1404. [CrossRef]
6. Meng, L.; Liu, X.; Liu, L.; Hong, Q.; Cheng, Y.; Gao, F.; Chen, J.; Zhang, Q.; Pan, C. Comparative investigation of the corrosion behavior and biocompatibility of the different chemical conversion coatings on the magnesium alloy surfaces. *Metals* **2022**, *12*, 1644. [CrossRef]
7. Han, Z.; Guo, H.; Wang, L.; Zhang, K.; Li, J.A. Composite coating prepared with ferulic acid to improve the corrosion resistance and blood compatibility of magnesium alloy. *Metals* **2022**, *12*, 545. [CrossRef]
8. Sheng, Y.; Hou, R.; Liu, C.; Xue, Z.; Zhang, K.; Li, J.; Guan, S. Tailoring of biodegradable magnesium alloy surface with Schiff base coating via electrostatic spraying for better corrosion resistance. *Metals* **2022**, *12*, 471. [CrossRef]

9. Hu, Q.; Liu, H.; Gao, F.; Yang, X.; Li, J.; Liu, R.; Liu, Z.; Wang, D. Investigation on blood compatibility of Cu/Ti metal coating prepared via various bias voltages and copper content. *Metals* **2022**, *12*, 435. [CrossRef]
10. Zhang, Z.; Song, Q.; Jin, Y.; Feng, Y.; Li, J.; Zhang, K. Advances in Schiff base and its coating on metal biomaterials—A review. *Metals* **2023**, *13*, 386. [CrossRef]

metals

Article

Mechanical Properties of LPBF-Built Titanium Lattice Structures—A Comparative Study of As-Built and Hot Isostatic Pressed Structures for Medical Implants

Benedikt Adelmann * and Ralf Hellmann

Applied Laser and Photonics Group, Faculty of Engineering, University of Applied Sciences Aschaffenburg, Wuerzburger Strasse 45, 63734 Aschaffenburg, Germany
* Correspondence: benedikt.adelmann@th-ab.de

Abstract: We compare different lattice structures with various elementary cell sizes built by laser powder bed fusion with and without hot isostatic pressing as post treatment. Cylindrical lattice structures are mechanically tested upon static and dynamic load in order to achieve high elasticity, high fracture strength and a high number of cycles to failure with respect to applications as medical implants. Evaluating the Young's modulus, a high stiffness for the body diagonal structure and a low fracture stress for the G-structure are measured. Hot isostatic pressing results in a higher Young's modulus and is ambiguous in terms of fractural stress. While samples without hot isostatic pressing reveal a shear fracture, the hot isostatic pressed samples have a high ductile area where the lattice layers are wrapped and pressed into the underlying layers without a fracture. Under dynamic load, the samples without hot isostatic pressing mostly are unable withstand 10^6 cycles at typical loads of the human body. Hot isostatic pressing has no significant influence on the strength at high loads and low cycle numbers, but at low loads all samples survived 10^6 cycles. As a consequence, dode-thick and rhombic dodecahedrons with 2 mm and 1.5 mm lattice size after hot isostatic pressing are recommended for medical implants because of the high elasticity, high fracture stress and high resistance against dynamic loads, which fulfill implant requirements.

Keywords: additive manufacturing; laser powder bed fusion; titanium; lattice structures; medical implant

Citation: Adelmann, B.; Hellmann, R. Mechanical Properties of LPBF-Built Titanium Lattice Structures—A Comparative Study of As-Built and Hot Isostatic Pressed Structures for Medical Implants. *Metals* **2022**, *12*, 2072. https://doi.org/10.3390/met12122072

Academic Editors: Changjiang Pan, Jingan Li and Pavel Krakhmalev

Received: 28 September 2022
Accepted: 23 November 2022
Published: 2 December 2022

1. Introduction

Life expectancy rises continuously worldwide, changing the age structure of the population pyramid. This socio-demographic change impacts social and commercial prospects and creates increasing and new demands for care and medical technologies. Next to the therapy of degenerative diseases of the central nervous system, preserving mobility is a key requirement that inevitably challenges the medical technology sector [1–3].

Laser powder bead fusion (LPBF), also denoted as selective laser melting [4], features the potential for improving medical implant design, in turn improving the mobility of invalid and older people. Specifically, medical grade titanium alloys are in favor for implants because of their superior biocompatibility [5–12]. Today, spinal implants are primarily manufactured by casting or milling. Though implant technology is quite advanced, milled implants still exhibit deficits, such as stress shielding, due to the high stiffness of implants as compared to natural tissue. Stress shielding results in a capacity underload of the protected natural bone, leading in a reduction in bone density and weakening of healthy bone tissue. Stress shielding appears when the inserted implant mechanically buffers the remaining tissue, destabilizing the balance of formation and degradation in bone tissue [13]. In order to circumvent stress shielding in LPBF-made implants, Fousová et al. built complex lattice structures of 316L stainless steel to copycat the mechanical elasticity of the trabecular human bone [14]. By producing porous materials using LPBF, stress shielding was reduced by 75% as compared to solid implants [15]. Printing porous implants of β-Ti-35Zr-28Nb using LPBF has led to a porosity of up to 83% and a Young's modulus in the range of trabecular bone [5]. In particular, LPBF allowed variance of Young's modulus by one order of magnitude, thus avoiding stress shielding [16]. Porous structures built from Ti16Nb by powder metallurgy using different amount of powders have shown mechanical properties similar to human bones [17]. LPBF even allows a mixture of powders like hydroxyapatite (a matrix for bones) and 316L stainless steel to produce a material with tensile strengths close to human bones [18]. Altogether, these results underline that different lattice structures can improve the elasticity and avoid possible stress shielding.

Beside such considerations as the Young's modulus and stress shielding, a medical implant has to last for several decades under dynamic loads, i.e., a high number of stress cycles. Therefore, the fatigue behavior of the implant is also of utmost importance and has thus been widely studied [6,19–21]. With respect to LPBF, the fatigue behavior is strongly negatively influenced by pores slightly underneath the surface [22] and from particles on the surface [23,24]. Because the position of such defects has a major influence, the fatigue behavior is hard to predict by fracture models [25,26].

With increasing defect size, the fatigue resistance decreases [27], so the LPBF process parameters have to be carefully adjusted. The microstructure during the solidification of the LPBF process also influences the fatigue behavior [28]. As LPBF parts do not adequately meet the requirements for implants, e.g., high service life and defect-free microstructure [29], post-treatment is compulsory according to ASTM F2924–14. In this respect, hot isostatic pressing (HIP) is well recognized for a reduction of inner pores and cavities as well as for an increase in density [30,31]. Studies have revealed that HIP treatment of Ti–6Al–2Zr–1Mo–1V results in a 15% increase in tensile strength and a 53% increase in ductility [32]. Moreover, hot isostatic pressing significantly reduces the residual stress in LPBF built parts. This typically results in an increase of fracture toughness and thus in a significantly stronger plastic area, as reported for aluminum alloys [33], stainless steel [34] and special alloys [35]. This in turn leads in titanium to an improved number of cycles to failure due to hot isostatic pressing [36,37]. In addition, unwanted anisotropy like build direction–dependent strength can be removed [38,39]. For Ti–6Al–2Zr–1Mo–1V, the tensile strength and ductility of the HIP specimens at 500 °C have been reported as 14.8% and 52.8% higher than those of their counterparts [32], respectively.

In addition to the often-used Ti-6Al-4V, other titanium-based alloys are under study to improve the mechanical properties of implants. For instance, Mo2C/Ti64 composites reveal a remarkable improvement of 23% higher hardness and 20% higher tensile strength by adding Mo2C [40]. Ti–16Nb–xZr alloys have been studied [41], showing advantages in terms of cost and easy manufacturability as compared to other titanium alloys. Compared to titanium, Ti-Nb alloys are characterized by a higher hardness and transverse rupture strength, while simultaneously the elastic modulus is decreased [42].

The underlying metallographic properties and processes during LPBF and hot isostatic pressing of medically relevant titanium alloys have been comprehensively studied in previous works. To guide the reader to further reading, we refer to [31,32,43], while in this contribution we focus on a comparative study of different lattice structures.

Because the fatigue behavior of LPBF built structures depends strongly on the build geometry [44], e.g., thin wall-based structures have a better fatigue resistance than lattice structures [45,46], each structure in combination with a thermal post process has to be tested individually towards fatigue behavior. Therefore, in this contribution, lattice structures for medical implants, adopted by the commonly utilized Materialise Magics® software (Ver. 25.3, Leuven, Belgium), are examined, and the influence of hot isostatic post processing on the strength, elasticity and fatigue behavior of these lattice structures is evaluated and discussed in terms of fulfillment of the requirements of the medical application.

With respect to geometrical restrictions, the size of typical intervertebral disc implants varies in the range of 5 mm to 11 mm in height, from 6 mm to 19 mm in width and from 12 mm to 18 mm in depth [47]. Footprints of implants vary from circles and triangles to rectangles, ovals, kidney shapes and curved geometries [47]. For the typical load, different typical postures have to be considered. Typical forces on the intervertebral disc are 0.7 kN when sitting, 1.9 kN when lifting 10 kg and 9 kN when lifting 50 kg [48]. Therefore, a movement that is performed once a day results after 30 years in about 10^4 movements. Assuming that a weight of 50 kg should not be lifted several times a day, for the load of 9 kN about 10^4 cycles is enough. At the same time, weights of 10 kg are lifted very often tens of times a day depending on the motion activity of a person, so an implant has to last 10^6 cycles. In everyday life, the typical internal pressure in an intervertebral disc is 2 MPa, and a symmetrically loaded intervertebral disc collapses at a load of 11 MPa. Therefore, we also want to prove whether LPBF built lattice structures fulfill these requirements.

2. Experiment

2.1. Materials

The material used for the experiments is Ti-6Al-4V Grade 23 (Heraeus, Hanau, Germany) with the material number 3.7165, which is one of the most often used titanium alloys for medical applications. According to the datasheet, the density of 4.43 g/cm^3 classifies the material as lightweight alloy. The material is characterized by a high strength at low density, an excellent corrosion resistance and high biocompatibility. Due to a low amount of interstitial iron and oxygen (below 0.3 wt%), the material is suitable for medical implants and dentistry implants. In detail, the alloy consists of 6 wt% aluminum and 4 wt% vanadium, which is balanced with titanium. The material is powder, with a grain size between 15 μm and 53 μm and the particles having a spherical shape. According to the datasheet, the tensile strength is 1280 MPa as built and 960 MPa after heat treatment. Further, the Young's modulus is given with 115 GPa as built and 125 GPa after heat treatment.

Regarding the high cooling rate during LPBF, Ti-6Al-4V represents an α-martensite after the additive manufacturing process. Upon hot isostatic pressing below the β-transus temperature of 980 °C, the crystalline structure of the metal is transformed to a mixture of α and β fractions, which are often lamellar shaped. The material then still includes proportions of pure α-martensite phases. For temperatures above the β-transus point during HIP treatment, large β-grains develop, which alter to a lamellar-shaped α–β mixture when cooling down after the heat treatment [43,49] and influence the mechanical properties towards higher cutting forces during milling and higher ductility [9,49].

2.2. Lattice Structures

To achieve the targeted high elasticity, 5 different lattice structures, chosen and adopted utilizing Materialise Magics® software, as shown in Figure 1, are studied. The body diagonal structure has a cubic elementary cell with round bars and rounded nodes on the intersection points. The diamond 30% density structure (hereafter named diamond) also consists of round bars, building a diamond elementary cell, but has no rounding on the intersection points. In addition, with round bars and no rounding on the intersection point, a rhombic dodecahedron elementary cell is built. The dode-thick structure is built from triangular bars without rounding at the intersection point. The G-structure has the same elementary cell as the body diagonals but uses triangular bars with no rounding at the intersection points. All structures are constructed using the "Materialize Magic Structures" toolbox.

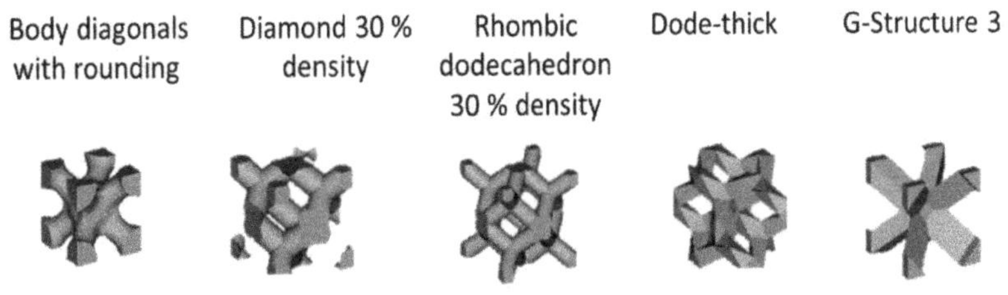

Figure 1. Scheme of different lattice structures.

2.3. LPBF Process

We employed a Lasertec 30 machine (DMG Mori Additive, Bielefeld, Germany) for the LPBF of Ti-6Al-4V (material number 3.7165), building cylinders with dimensions of 10 mm and 15 mm in diameter and height, in accordance with DIN 50106 "Testing of metallic materials—Compression test at room temperature". This size is in the range of typical intervertebral disc implants. The volume was filled by five different lattice structures using the software Materialize Magics, as shown in Figure 1. In total, three variations of the test specimen were created per unit cell. The size of the elementary cell was varied between 1 mm, 1.5 mm and 2 mm. The layer height during the LPBF process was 50 μm. The employed laser machine parameters were a scan velocity of 0.74 m/s, a laser power of 220 W and hatch distance of 0.10 mm. The scan strategy was to fill the necessary areas with lines, whose angle changes from layer to layer. After this infill, the borderlines of the edges were scanned by the laser beam. For the built parts, this parameter combination led to a comparatively low surface roughness R_a of about 10 μm and a high density.

2.4. Hot Isostatic Pressing

The machine used for hot isostatic pressing was a Q15L 2070-1400M from Quintus Technologies AB (Vasteras, Sweden). The system provides a maximum pressure of 200 MPa and a maximal temperature of 1400 °C under argon atmosphere. The process chamber is 300 mm in height and 170 mm in diameter, which allows a maximum load of 40 kg. The temperature control has an accuracy of ±10 °C and provides a maximum heating rate of 30 °C/min.

Hot isostatic pressing cycles were performed in accordance with ASTM F2924–14, which translates to treatment with 100 MPa pressure at a temperature of 925 °C for a 3 h period. In detail, the HIP process was divided into four steps. In the first step, the process chamber is heated to 890 °C, with a heating rate of 10°/min and simultaneously the pressure is increased to 100 MPa. During the second step, the heating rate is reduced to 3°/min up to a temperature of 925 °C, to avoid overshooting. In a third step, the temperature (925 °C) and pressure (100 MPa) are held for 3 h before the chamber cools down slowly due to natural heat conduction in the fourth step, which takes several hours.

To avoid confusion, in this contribution, samples undergoing HIP are named as post HIP, while samples measured before the HIP process are referred to as pre-HIP.

2.5. Mechanical Testing

Compressive strength measurements were performed with a universal testing machine Shimadzu Autograph AG-X plus (Shimadzu, Kyoto, Japan), using a speed of 1 mm/min. The compressive fatigue behavior tests were carried out on the dynamic tensile test rig STEPLab UD020 (STEPLab, Resana, Italy). A sinusoidal stress–time function with a frequency of 10 Hz and with a force between 0 N and the nominal value was used. The specimens were run through the loading case until either a break occurred or 10^6 cycles were reached. A total of three measurements were made for each load case. The resulting cycles to failure were recorded in a Wöhler curve (S–N curve). For all mechanical tests, three samples per group were used in order to achieve a reliable value and its deviation.

3. Results

3.1. Print Characteristics

To characterize printing, the relative density of the lattice structures was determined, and cross-section images were taken. To determine the relative density, the surrounding volume of the cylinders was measured, and the samples were weighed. The weight was divided by the volume and the density of solid Ti-6Al-4V. The results are shown in Figure 2 for the five lattice structures with the three elementary cell sizes under study. The results reveal decreasing densities with increasing elementary cell sizes for all cell types except the G-structure. A possible reason for this is that the size of the rods is smaller to the minimum producible size of the laser machine. As a consequence, the rods are produced larger than designed, and the density increases. For larger elementary cell size, these rods are designed larger, and the oversizing of the machine is reduced. A comparison between the cell types reveals the highest density for body diagonals and the rhombic dodecahedron and the lowest for the G-structure.

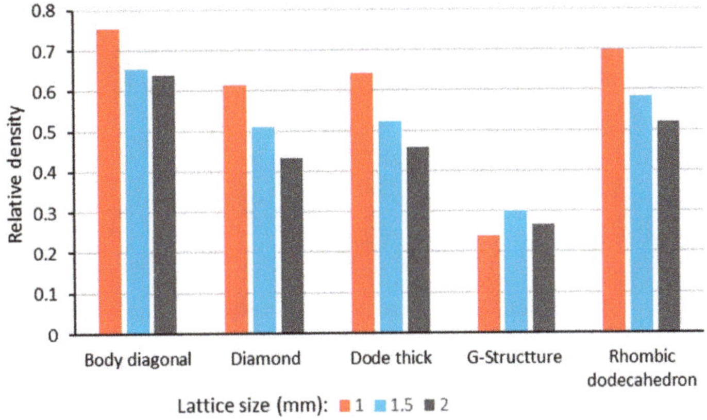

Figure 2. Relative density of different lattice structures.

To study porosity, cross-section images were taken from both pre- and post-HIP specimens (cf. Figure 3. As depicted in Figure 3a, the porosity is quite low, but there are pores with sizes up to about 50 μm. Several further cross-section images reveal that pores are distributed in many areas all over the sample. After the HIP process (Figure 3b), no pores were visible. When analyzing several cross-section images, it was observed that after HIP, the size and number of pores were reduced as compared to pre-HIP.

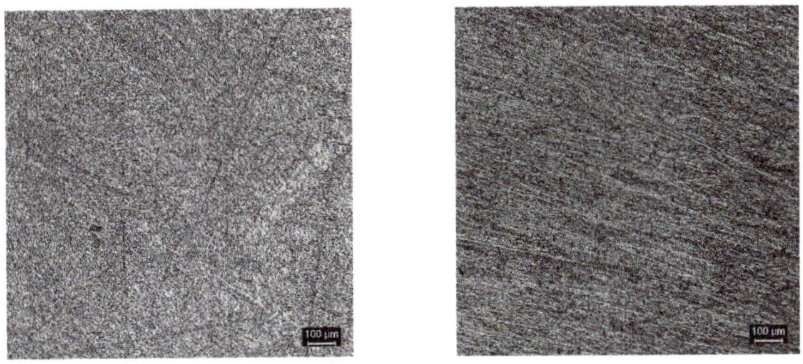

(**a**) (**b**)

Figure 3. Cross-section images (**a**) without HIP and (**b**) after HIP.

3.2. Stress–Strain Behavior

To exemplify the stress–strain properties of the evaluated lattice structures, Figure 4 summarizes the stress–strain behavior of the dode-thick structure for different lattice sizes before and after hot isostatic pressing. It should be noted that the stress–strain behavior of the other lattice structures under investigation reveals similar qualitative characteristics and quantitative results such as the Young's modulus and fracture stress, the results of which will be shown in Section 3.3 (Figures 6 and 7).

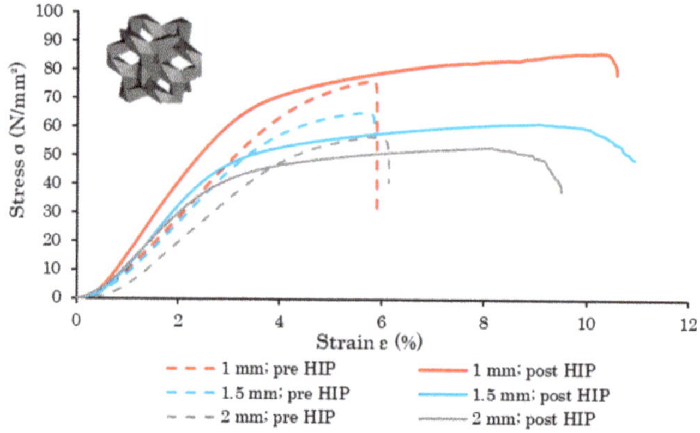

Figure 4. Stress–strain diagram of a dode-thick lattice structure.

As the results in Figure 4 show, after hot isostatic pressing, the elasticity reveals, in accordance with Ref. [49], a higher stiffness as compared to the pre-HIP state for all structure sizes. With decreasing elementary cell size, both stiffness and fracture strength increase for both pre-HIP and post-HIP structures, a trend that was previously shown for body-centered cubic Ti-6Al-4V lattices (without hot isostatic pressing) by Maskery et al. [50]. In addition, the diagram reveals a significantly higher strain of the post-HIP samples, caused by an extended range of plastic deformation. This agrees with the literature results, revealing a substantial ductility improvement for SLM-built titanium samples [31], which in turn is associated with the characteristic fracture and compression behavior of the pre-HIP and post-HIP structures. Figure 5 depicts the samples after the compression test, showing that the pre-HIP samples in Figure 5a are characterized by a shear fracture through the complete sample, as is typical for brittle structures. Contrary to this, the post-HIP samples in Figure 5b do not show any fracture rather than a strong deformation of the entire lattice structure, with layers being pressed into the underlying layers. This can be attributed to a fracture strain increase of up to 14% after thermal post treatment, while at the same time the yield stress decreases, as was reported for Ti-6Al-4V by Vrancken et al. [43]. Similar experiments for nickel-based superalloys show a 410% higher ductility after a HIP treatment, followed by a solid solution heat treatment [51].

(**a**) (**b**)

Figure 5. Failure of dode-thick lattice structures with 10 mm diameter under static load before HIP (**a**) and after HIP process (**b**).

3.3. Elasticity and Strength

To identify the most suitable lattice structure and elementary cell size in terms of high elasticity and to avoid stress shielding and damage under a single high load, as for lifting a very heavy object, the Young's modulus and the fractures stress under static load were determined. Figure 6 gives an overview of the Young's modulus for the different lattice structures and elementary cell sizes under study, as well as for specimens without and with hot isostatic pressing. For all specimens, hot isostatic pressing leads to an increasing stiffness in the range between 10% and 20%. Furthermore, for all lattice structures except the G-structure, Young's modulus increases with decreasing elementary cell size. A disadvantage of small elementary cell sizes is a higher standard deviation, which can be attributed to the smaller web thickness, for which SLM's typical defects, such as pores, have a higher influence. In addition, comparing the lattice structures amongst each other, the G-structure exhibits the lowest, whereas the Body diagonal structure exhibits the highest Young's modulus. Diamond and dode-thick structures reveal similar elasticity, while rhombic dodecahedrons appear slightly stiffer. In this context, the Young's modulus is dependent on the relative density (see Figure 2), while the Body diagonal has the highest relative density and the highest Young's modulus and the G-structure is lowest in both categories. For diamond and dode-thick structures, the Young's modulus and the relative density are similar, and the rhombic dodecahedron is slightly higher in both density and Young's modulus.

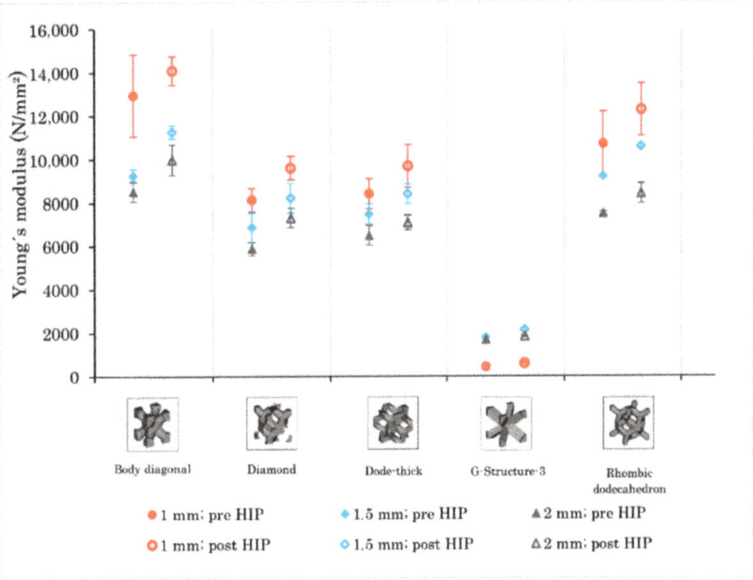

Figure 6. Young's modulus of different lattice structures as a function of lattice size and HIP.

To avoid the fracture of an implant during a high load, the fracture stress should be as high as possible. The fracture stress for different lattice structures and elementary cell sizes is summarized in Figure 7. For all lattice structures, a high fracture strength is accompanied with a high Young's modulus (see Figure 6). In contrast to the Young's modulus, hot isostatic pressing has no definite impact on the fracture stress, showing only a minor influence in both directions of higher or lower fracture stress for the different lattice structures and elementary cell sizes. In contrast to these measurements, in the literature, HIP treatment of Ti–6Al–2Zr–1Mo–1V is reported to result in a 15% increase in tensile strength [32]. However, the influence of the elementary cell size reveals an increasing fracture stress with decreasing cell size, except for the G-structure, a behavior similarly found for the Young's modulus. The cell type with the highest fracture stress is the Body diagonal, while the G-structure has the lowest stress, which in turn is roughly in accordance with the relative density (see Figure 2).

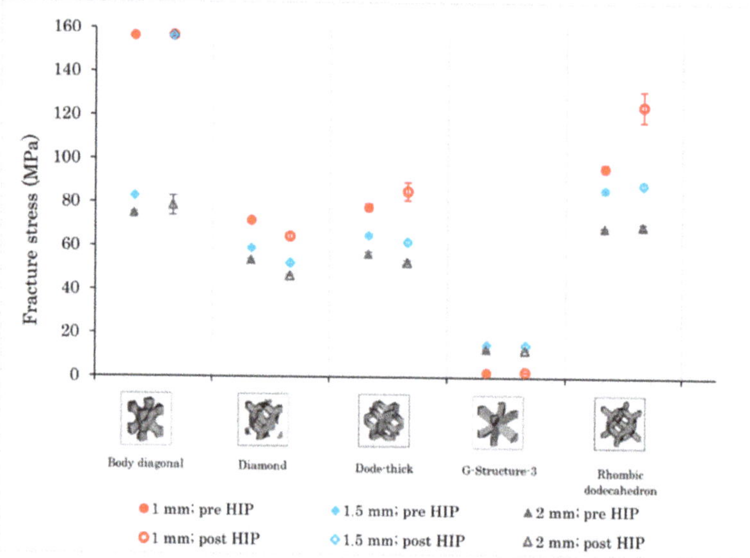

Figure 7. Fracture stress of different lattice structures as a function of lattice size and HIP.

With regard to the desired application and the required mechanical properties, a fracture stress above 11 MPa and a low Young's modulus are necessary. Based on the results discussed above, therefore, certain structures and parameter combinations can be excluded. For instance, the fracture stress of the G-structure depends on the cell size in the same range or below the collapse load of a natural intervertebral disc of 11 MPa [48]. Hence, the G-structure can be omitted. All structures with an elementary cell size of 1 mm can be excluded because of the high Young's modulus and the high deviation. Because of its high Young's modulus, the Body diagonal structure and the rhombic dodecahedron with 1.5 mm elementary cell size are also disqualified. The remaining structures and parameter combinations are characterized by a sufficiently high fracture stress and a high elasticity below 7.500 N/mm^2 in the pre-HIP status and below 8.500 N/mm^2 in the post-HIP status.

3.4. Fatigue Behavior

To determine the fatigue behavior, Wöhler tests were carried out, with loads related to the application as an intervertebral disc implant. Thus, three measuring points were defined: 1.9 kN, which corresponds to the compressive force in the lumbar intervertebral disc when lifting 10 kg; 9 kN, corresponding to the maximum compressive force occurring in the lumbar intervertebral disc during the stooped lifting of more than 50 kg; and 13.5 kN, which includes a safety factor of 1.5 times the maximum load. A specimen was considered suitable if its fatigue resistance in the load range of 1.9 kN withstood the load for more than 10^6 cycles, which corresponds to about 100 lifts per day for 30 years. In case of a failure of a sample, the cycles to failure were recorded.

The results for the pre-HIP samples of the remaining lattice structures, i.e., after the exclusion of certain selected structures discussed in the previous section, are depicted in a Wöhler curve in Figure 8. Only the test specimen with dode-thick lattice and 1.5 mm elementary cell size fulfilled the required fatigue strength under a load of 1.9 kN. All other specimens displayed anterior fracture. Furthermore, it can be stated that the test specimen rhombic dodecahedron with 2 mm elementary cell size exhibited the highest strength at high loads and a low number of cycles before fracture but showed a disproportionately high decrease of the stress amplitude to failure at a high number of cycles. The diamond-30-percent-relative-density structure with an elementary cell size of 2 mm, on the other hand, exhibited the lowest strength at all numbers of cycles. Diamond-30 and dode-thick structures exhibited comparable densities; any differences might be attributed to the number of the struts of these lattice structures, which determine the Maxwell criterion of cell-based lattice structures, and thus the dominating load response type, and to the geometry of the struts (round bars for Diamond-30 and triangular bars for dode-thick; see Section 2.2). In addition, such geometries may exhibit differences in microstructure quality.

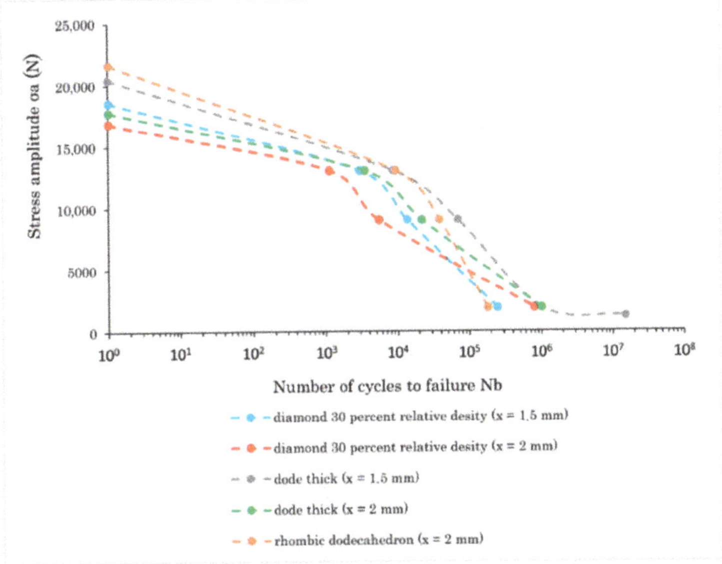

Figure 8. Fatigue behavior of different lattice structures pre-HIP.

Overall, the pre-HIP structures revealed a major lack of fatigue strength at low loads and a high number of cycles. This is attributed to defects like pores and micro cracks during the LPBF process [22,23], which grow and propagate slowly in volume with an increasing number of cycles [24].

In addition, the post-HIP samples were dynamically tested to determine the influence of hot isostatic pressing on the fatigue behavior. The results of the fatigue tests are shown in Figure 9 and reveal no major differences to Figure 8, i.e., to the pre-HIP status, at one cycle and also no clear influence at loads of 13 kN and 9 kN. Looking at the low load part of the curve with 1.9 kN load, it can be seen that all specimens withstand the load for 10^6 cycles and the 1.2 kN load for 10^7 cycles. The dynamic load capacity increases significantly at a low load and a high number of cycles in comparison to the specimen pre-HIP.

The optical examination of the specimen shows a similar behavior to the static compression tests, as shown in Figure 5. While the pre-HIP components tend to show shear fracture as reported in the literature [52], the post-HIP components show an increased tendency to severe deformations and displacements of the lattice planes. This is particularly evident in the fact that some post-HIP components reveal a strong deformation of lattice planes and, in some cases, shear fractures of individual areas, but the component as a whole does not show total failure similar to Figure 5a.

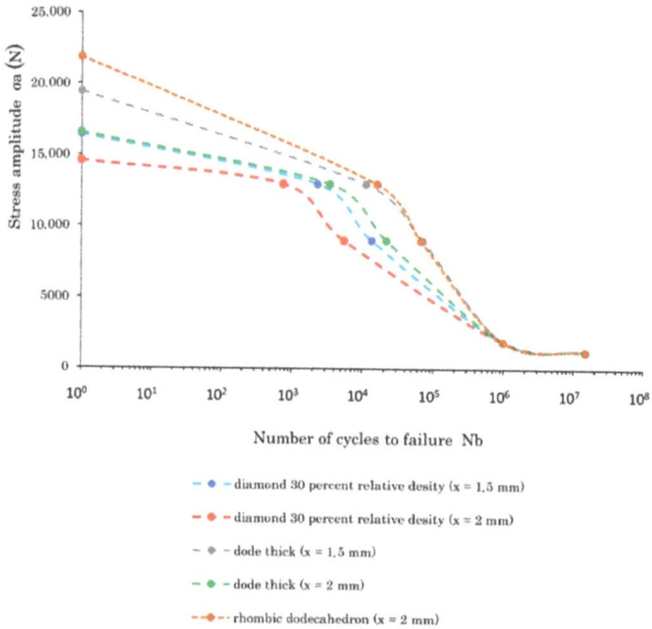

Figure 9. Fatigue behavior of different lattice structures post-HIP.

4. Discussion

In both the static and the dynamic compression tests, a significant influence of hot isostatic pressing on the examined components is found. In the static compression tests, it is particularly noticeable that the plastic range of the stress–strain diagram is significantly larger for post-HIP specimens as compared to the pre-HIP status. This agrees with the literature results, revealing a higher fracture strain after heat treatment [43]. This can be explained by the reduction of residual stresses during hot isostatic pressing, which results in a larger elongation capacity. In addition, an ambiguous influence on the fracture stress is measured, which can be attributed to the reduction of residual stress after HIP, which decreases the fracture stress. Without hot isostatic pressing (pre-HIP status), these stresses counteract the forces acting from outside and thus require a higher force to destroy the component; after hot isostatic pressing (post HIP status) no internal stresses counteract the forces from the outside, e.g., its fracture occurs at a lower stress than pre HIP [53]. Moreover, a reduction of pores and micro cracks during the HIP process increases the fracture stress, which counteracts the loss of internal stresses, and the influence of the HIP process results are ambiguous as measured.

The examination of the dynamic load revealed that only one pre-HIP sample fulfills the requirement of withstanding 10^6 cycles (dode-thick with 1.5 mm cell size). After heat treatment, however, the strength of the samples increases significantly at a low load and high number of cycles. At a high number of cycles, cracking occurs from defects such as pores or micro cracks [25,46]. Pre-HIP components have a large number of defects inside or on the surface of the component, while the inner defects are reduced in size and number by hot isostatic pressing [31]. So, the crack growth, as occurring pre HIP, can consequently be prevented or significantly delayed. In the case of high stress and low cycle numbers, on the other hand, the behavior is different. However, hot isostatic pressing, at least with the prescribed process parameters, leads to a coarsening of the microstructure test specimen post HIP and can have a similar or lower load capacity compared to a test specimen pre HIP at high stress and low number of cycles.

For the selection of a suitable unit cell and thus for the optimization of the mechanical properties of an intervertebral disc implant, it is essential to consider the results of the static and dynamic tests. The static compression tests can be used to determine these specimens, which exhibit a sufficiently high fracture stress, and simultaneously the Young's modulus is kept comparatively low. The dynamic investigations post HIP initially suggest that all components exhibit a sufficiently high fatigue strength. However, the optical examination reveals that although there is no total failure of the specimens, most of the specimens show deformations and/or partial fractures. Only the rhombic dodecahedron with 2 mm cell size and dode-thick specimen with 1.5 mm cell size do not show any detectable defects after 10^6 cycles.

Detailed examination of the static compression tests of the most promising lattice types in Figure 6 reveals that the dode-thick specimen with an elementary cell size of 1.5 mm has the same Young's modulus as the rhombic specimen. In addition, the dynamic compression tests show that this elementary cell offers sufficiently high strength of more than 70,000 cycles, even at the typically occurring maximum force of 9 kN. Furthermore, sufficient fatigue strength at 1.9 kN is already present before hot isostatic pressing. The dode-thick elementary cell proves to be particularly promising in comparison with the rhombic dodecahedron cell, especially for the everyday load cases of 1.9 kN or 9 kN (Table 1).

Table 1. Static and dynamic properties of the most suitable lattice structures.

Geometry	Status	Young's Modulus (N/mm^2)	Fracture Stress (MPa)	Cycles at Fracture 1300 N	9000 N	1900 N
Rhombic dodecahedron (x = 2 mm)	Pre-HIP	7477	68	9169	37,602	176,156
	Post-HIP	8406	69	15,963	66,452	>10^6
Dode-thick (x = 1.5 mm)	Pre-HIP	7476	64	8605	68,790	>10^6
	Post-HIP	8395	61	11,021	70,072	>10^6

In summary, it can be stated that the mechanical optimization of the component in order to achieve the lowest possible Young's modulus combined with a high fatigue strength can be realized by selecting a suitable lattice structure. In comparison to massive titanium with a Young's modulus of 110 kN/mm^2, the Young's modulus is reduced by using lattice structures of about 93%. Taking into account the results obtained, the dode-thick specimen with an elementary cell size of 1.5 mm proves to be the most promising for this purpose.

5. Conclusions

We compared the mechanical load properties of five Ti-6Al-4V lattice structure types with different elementary cell sizes built by laser powder bed fusion. In addition, the impact of hot isostatic pressing on the load capabilities under static and dynamic compression was studied. For comparison, the Young's modulus, the fracture stress and the cycles to failure were measured in order to evaluate the potential use of the lattice structures as medical intervertebral disc implants. In the static compression tests, the post HIP samples reveal a much higher strain, higher Young's modulus and a similar fracture stress. While samples without hot isostatic pressing reveal a shear fracture, the hot isostatic pressed samples have a high ductile area, where the lattice layers are wrapped and pressed into the underlying layers without fracture. In the dynamic tests, in the high load area no significant influence of the HIP process is measured, while in the high cycle low load regime the pre-HIP samples break, but the post samples survive the test. Overall, dode-thick and rhombic dodecahedron structures with 2 mm and 1.5 mm elementary cell size reveal no damage after 10^6 cycles and are recommended for the application due to their low Young's modulus.

Author Contributions: Conceptualization, B.A. and R.H.; methodology, B.A. and R.H.; software, B.A; validation, B.A.; formal analysis, B.A.; investigation, B.A.; resources, R.H.; data curation, B.A.; writing—original draft preparation, B.A. and R.H.; writing—review and editing, B.A. and R.H.; visualization, B.A.; supervision, R.H.; project administration, R.H. All authors have read and agreed to the published version of the manuscript.

Funding: This research received no funding.

Data Availability Statement: Not applicable.

Acknowledgments: We thank Melanie Abb for fruitful support and participating in conducting the experimental work.

Conflicts of Interest: The authors declare no conflict of interest.

References

1. Ferrucci, L.; Cooper, R.; Shardell, M.; Simonsick, E.M.; Schrack, J.A.; Kuh, D. Age-Related Change in Mobility: Perspectives From Life Course Epidemiology and Geroscience. *J. Gerontol. Ser. A Biol. Sci. Med. Sci.* **2016**, *71*, 1184–1194. [CrossRef] [PubMed]
2. Keehan, S.P.; Stone, D.A.; Poisal, J.A.; Cuckler, G.A.; Sisko, A.M.; Smith, S.D.; Madison, A.J.; Wolfe, C.J.; Lizonitz, J.M. National Health Expenditure Projections, 2016–2025: Price Increases, Aging Push Sector To 20 Percent Of Economy. *Health Aff. (Proj. Hope)* **2017**, *36*, 553–563. [CrossRef]
3. McInnes, L. Importance of maintaining mobility to elderly health. *Aging Health* **2011**, *7*, 165–167. [CrossRef]
4. *ISO/ASTM 52900:2015, ISO/TC 261*; Additive Manufacturing—General Principles—Terminology. International Organization for Standardization: Geneva, Switzerland, 2015.
5. Li, Y.; Ding, Y.; Munir, K.; Lin, J.; Brandt, M.; Atrens, A.; Xiao, Y.; Kanwar, J.R.; Wen, C. Novel β-Ti35Zr28Nb alloy scaffolds manufactured using selective laser melting for bone implant applications. *Acta Biomater.* **2019**, *87*, 273–284. [CrossRef] [PubMed]
6. Niinomi, M.; Nakai, M. Titanium-Based Biomaterials for Preventing Stress Shielding between Implant Devices and Bone. *Int. J. Biomater.* **2011**, *2011*, 836587. [CrossRef]
7. Wu, Z.; Dai, Y.; Luo, J.; Ji, X.; Xie, Z.; Li, L.; Xie, X. Physiochemical and biological evaluation of SLM-manufactured Ti-10Ta-2Nb-2Zr alloy for biomedical implant applications. *Biomed. Mater.* **2020**, *15*, 45017. [CrossRef]
8. Iwaya, Y.; Machigashira, M.; Kanbara, K.; Miyamoto, M.; Noguchi, K.; Izumi, Y.; Ban, S. Surface properties and biocompatibility of acid-etched titanium. *Dent. Mater. J.* **2008**, *27*, 415–421. [CrossRef]
9. Eisenbarth, E.; Velten, D.; Müller, M.; Thull, R.; Breme, J. Biocompatibility of beta-stabilizing elements of titanium alloys. *Biomaterials* **2004**, *25*, 5705–5713. [CrossRef]
10. Adelmann, B.; Abb, M.; Hellmann, R. Comparative study of cell growth and cellular adhesion on Ti-6Al-4V surfaces made by Selective Laser Melting followed by different surface post processing steps. *IOP Conf. Ser. Mater. Sci. Eng.* **2021**, *1135*, 12028. [CrossRef]
11. Sidambe, A.T. Biocompatibility of Advanced Manufactured Titanium Implants-A Review. *Materials* **2014**, *7*, 8168–8188. [CrossRef]
12. Matena, J.; Petersen, S.; Gieseke, M.; Kampmann, A.; Teske, M.; Beyerbach, M.; Escobar, H.M.; Haferkamp, H.; Gellrich, N.-C.; Nolte, I. SLM produced porous titanium implant improvements for enhanced vascularization and osteoblast seeding. *Int. J. Mol. Sci.* **2015**, *16*, 7478–7492. [CrossRef] [PubMed]
13. Edelmann, A.; Dubis, M.; Hellmann, R. Selective Laser Melting of Patient Individualized Osteosynthesis Plates-Digital to Physical Process Chain. *Materials* **2020**, *13*, 5786. [CrossRef]
14. Fousová, M.; Kubásek, J.; Vojtěch, D.; Fojt, J.; Čapek, J. 3D printed porous stainless steel for potential use in medicine. *IOP Conf. Ser. Mater. Sci. Eng.* **2017**, *179*, 12025. [CrossRef]
15. Arabnejad, S.; Johnston, B.; Tanzer, M.; Pasini, D. Fully porous 3D printed titanium femoral stem to reduce stress-shielding following total hip arthroplasty. *J. Orthop. Res. Off. Publ. Orthop. Res. Soc.* **2017**, *35*, 1774–1783. [CrossRef] [PubMed]
16. Leordean, D.; Dudescu, C.; Marcu, T.; Berce, P.; Balc, N. Customized implants with specific properties, made by selective laser melting. *Rapid Prototyp. J.* **2015**, *21*, 98–104. [CrossRef]
17. Yılmaz, E.; Gökçe, A.; Findik, F.; Gulsoy, H.O.; İyibilgin, O. Mechanical properties and electrochemical behavior of porous Ti-Nb biomaterials. *J. Mech. Behav. Biomed. Mater.* **2018**, *87*, 59–67. [CrossRef]
18. Hao, L.; Dadbakhsh, S.; Seaman, O.; Felstead, M. Selective laser melting of a stainless steel and hydroxyapatite composite for load-bearing implant development. *J. Mater. Process. Technol.* **2009**, *209*, 5793–5801. [CrossRef]
19. Chan, K.S.; Koike, M.; Mason, R.L.; Okabe, T. Fatigue Life of Titanium Alloys Fabricated by Additive Layer Manufacturing Techniques for Dental Implants. *Metall. Mater. Trans. A* **2013**, *44*, 1010–1022. [CrossRef]
20. Okazaki, Y.; Ito, Y.; Kyo, K.; Tateishi, T. Corrosion resistance and corrosion fatigue strength of new titanium alloys for medical implants without V and Al. *Mater. Sci. Eng. A* **1996**, *213*, 138–147. [CrossRef]
21. Shemtov-Yona, K.; Rittel, D. Fatigue of Dental Implants: Facts and Fallacies. *Dent. J.* **2016**, *4*, 16. [CrossRef]
22. Glodez, S.; Klemenc, J.; Zupanic, F.; Vesenjak, M. High-cycle fatigue and fracture behaviours of SLM AlSi10Mg alloy. *Trans. Nonferrous Met. Soc. China* **2020**, *30*, 2577–2589. [CrossRef]

23. Koutiri, I.; Pessard, E.; Peyre, P.; Amlou, O.; de Terris, T. Influence of SLM process parameters on the surface finish, porosity rate and fatigue behavior of as-built Inconel 625 parts. *J. Mater. Process. Technol.* **2018**, *255*, 536–546. [CrossRef]
24. Kan, W.H.; Chiu, L.N.S.; Lim, C.V.S.; Zhu, Y.; Tian, Y.; Jiang, D.; Huang, A. A critical review on the effects of process-induced porosity on the mechanical properties of alloys fabricated by laser powder bed fusion. *J. Mater. Sci.* **2022**, *57*, 9818–9865. [CrossRef]
25. Leuders, S.; Vollmer, M.; Brenne, F.; Tröster, T.; Niendorf, T. Fatigue Strength Prediction for Titanium Alloy TiAl6V4 Manufactured by Selective Laser Melting. *Metall. Mater. Trans. A* **2015**, *46*, 3816–3823. [CrossRef]
26. Ferreira, F.F.; Neto, D.M.; Jesus, J.S.; Prates, P.A.; Antunes, F.V. Numerical Prediction of the Fatigue Crack Growth Rate in SLM Ti-6Al-4V Based on Crack Tip Plastic Strain. *Metals* **2020**, *10*, 1133. [CrossRef]
27. Ngnekou, J.N.D.; Nadot, Y.; Henaff, G.; Nicolai, J.; Ridosz, L. Influence of defect size on the fatigue resistance of AlSi10Mg alloy elaborated by selective laser melting (SLM). *Procedia Struct. Integr.* **2017**, *7*, 75–83. [CrossRef]
28. Biffi, C.A.; Fiocchi, J.; Bassani, P.; Paolino, D.S.; Tridello, A.; Chiandussi, G.; Rossetto, M.; Tuissi, A. Microstructure and preliminary fatigue analysis on AlSi10Mg samples manufactured by SLM. *Procedia Struct. Integr.* **2017**, *7*, 50–57. [CrossRef]
29. Yadav, P.; Rigo, O.; Arvieu, C.; Le Guen, E.; Lacoste, E. In Situ Monitoring Systems of The SLM Process: On the Need to Develop Machine Learning Models for Data Processing. *Crystals* **2020**, *10*, 524. [CrossRef]
30. Tillmann, W.; Schaak, C.; Nellesen, J.; Schaper, M.; Aydinöz, M.E.; Hoyer, K.-P. Hot isostatic pressing of IN718 components manufactured by selective laser melting. *Addit. Manuf.* **2017**, *13*, 93–102. [CrossRef]
31. Jamshidi, P.; Aristizabal, M.; Kong, W.; Villapun, V.; Cox, S.C.; Grover, L.M.; Attallah, M.M. Selective Laser Melting of Ti-6Al-4V: The Impact of Post-processing on the Tensile, Fatigue and Biological Properties for Medical Implant Applications. *Materials* **2020**, *13*, 2813. [CrossRef]
32. Cai, C.; Gao, X.; Teng, Q.; Kiran, R.; Liu, J.; Wei, Q.; Shi, Y. Hot isostatic pressing of a near α-Ti alloy: Temperature optimization, microstructural evolution and mechanical performance evaluation. *Mater. Sci. Eng. A* **2021**, *802*, 140426. [CrossRef]
33. Aboulkhair, N.T.; Maskery, I.; Tuck, C.; Ashcroft, I.; Everitt, N.M. The microstructure and mechanical properties of selectively laser melted AlSi10Mg: The effect of a conventional T6-like heat treatment. *Mater. Sci. Eng. A* **2016**, *667*, 139–146. [CrossRef]
34. Sistiaga, M.L.M.; Hautfenne, S.N.C.; van Humbeeck, J. Effect of Heat Treatment Of 316L Stainless Steel Produced by Selective Laser Melting (SLM). In Proceedings of the 27th Annual International Solid Freeform Fabrication Symposium—An Additive Manufacturing Conference, Austin, TX, USA, 8–10 August 2016; University of Texas: Austin, TX, USA, 2016; pp. 558–565.
35. Spierings, A.B.; Dawson, K.; Kern, K.; Palm, F.; Wegener, K. SLM-processed Sc- and Zr- modified Al-Mg alloy: Mechanical properties and microstructural effects of heat treatment. *Mater. Sci. Eng. A* **2017**, *701*, 264–273. [CrossRef]
36. van Hooreweder, B.; Kruth, J.-P. Advanced fatigue analysis of metal lattice structures produced by Selective Laser Melting. *CIRP Ann.* **2017**, *66*, 221–224. [CrossRef]
37. Greitemeier, D.; Palm, F.; Syassen, F.; Melz, T. Fatigue performance of additive manufactured TiAl6V4 using electron and laser beam melting. *Int. J. Fatigue* **2017**, *94*, 211–217. [CrossRef]
38. Rans, C.; Michielssen, J.; Walker, M.; Wang, W.; Hoen-Velterop, L. Beyond the orthogonal: On the influence of build orientation on fatigue crack growth in SLM Ti-6Al-4V. *Int. J. Fatigue* **2018**, *116*, 344–354. [CrossRef]
39. Sangid, M.D.; Book, T.A.; Naragani, D.; Rotella, J.; Ravi, P.; Finch, A.; Kenesei, P.; Park, J.-S.; Sharma, H.; Almer, J.; et al. Role of heat treatment and build orientation in the microstructure sensitive deformation characteristics of IN718 produced via SLM additive manufacturing. *Addit. Manuf.* **2018**, *22*, 479–496. [CrossRef]
40. Cai, C.; Qiu, J.C.D.; Shian, T.W.; Han, C.; Liu, T.; Kong, L.B.; Srikanth, N.; Sun, C.-N.; Zhou, K. Laser powder bed fusion of Mo2C/Ti-6Al-4V composites with alternately laminated α'/β phases for enhanced mechanical properties. *Addit. Manuf.* **2021**, *46*, 102134. [CrossRef]
41. Yılmaz, E.; Gökçe, A.; Findik, F.; Gulsoy, H.Ö. Assessment of Ti–16Nb–xZr alloys produced via PIM for implant applications. *J. Therm. Anal. Calorim.* **2018**, *134*, 7–14. [CrossRef]
42. Yılmaz, E.; Gökçe, A.; Findik, F.; Gulsoy, H. Metallurgical properties and biomimetic HA deposition performance of Ti-Nb PIM alloys. *J. Alloys Compd.* **2018**, *746*, 301–313. [CrossRef]
43. Vrancken, B.; Thijs, L.; Kruth, J.-P.; van Humbeeck, J. Heat treatment of Ti6Al4V produced by Selective Laser Melting: Microstructure and mechanical properties. *J. Alloys Compd.* **2012**, *541*, 177–185. [CrossRef]
44. Agius, D.; Kourousis, K.; Wallbrink, C. A Review of the As-Built SLM Ti-6Al-4V Mechanical Properties towards Achieving Fatigue Resistant Designs. *Metals* **2018**, *8*, 75. [CrossRef]
45. Speirs, M.; van Hooreweder, B.; van Humbeeck, J.; Kruth, J.-P. Fatigue behaviour of NiTi shape memory alloy scaffolds produced by SLM, a unit cell design comparison. *J. Mech. Behav. Biomed. Mater.* **2017**, *70*, 53–59. [CrossRef] [PubMed]
46. Refai, K.; Brugger, C.; Montemurro, M.; Saintier, N. An experimental and numerical study of the high cycle multiaxial fatigue strength of titanium lattice structures produced by Selective Laser Melting (SLM). *Int. J. Fatigue* **2020**, *138*, 105623. [CrossRef]
47. Fritsch, E.W.; Pitzen, T. Zervikale Bandscheibenprothesen. *Der. Orthop.* **2006**, *35*, 347–359. [CrossRef]
48. Wintermantel, E.; Ha, S.-W. (Eds.) *Medizintechnik: Life Science Engineering*, 5th ed.; Springer: Berlin/Heidelberg, Germany, 2009.
49. Ahmadi, M.; Karpat, Y.; Acar, O.; Kalay, Y.E. Microstructure effects on process outputs in micro scale milling of heat treated Ti6Al4V titanium alloys. *J. Mater. Process. Technol.* **2018**, *252*, 333–347. [CrossRef]
50. Maskery, I.; Aremu, A.O.; Simonelli, M.; Tuck, C.; Wildman, R.D.; Ashcroft, I.A.; Hague, R. Mechanical Properties of Ti-6Al-4V Selectively Laser Melted Parts with Body-Centred-Cubic Lattices of Varying cell size. *Exp. Mech.* **2015**, *55*, 1261–1272. [CrossRef]

51. Sun, S.; Teng, Q.; Xie, Y.; Liu, T.; Ma, R.; Bai, J.; Cai, C.; Wei, Q. Two-step heat treatment for laser powder bed fusion of a nickel-based superalloy with simultaneously enhanced tensile strength and ductility. *Addit. Manuf.* **2021**, *46*, 102168. [CrossRef]
52. Mazur, M.; Leary, M.; Sun, S.; Vcelka, M.; Shidid, D.; Brandt, M. Deformation and failure behaviour of Ti-6Al-4V lattice structures manufactured by selective laser melting (SLM). *Int. J. Adv. Manuf. Technol.* **2015**, *84*, 1391–1411. [CrossRef]
53. Kumar, P.; Ramamurty, U. Microstructural optimization through heat treatment for enhancing the fracture toughness and fatigue crack growth resistance of selective laser melted Ti 6Al 4V alloy. *Acta Mater.* **2019**, *169*, 45–59. [CrossRef]

metals

Article

Biomimetic Design of Fatigue-Testing Fixture for Artificial Cervical Disc Prostheses

Xuejin Cheng [1,2], Jia Bai [3] and Tao Wang [1,*]

1 College of Materials Science and Technology, Nanjing University of Aeronautics and Astronautics, Nanjing 211106, China
2 Faculty of Mechanical and Material Engineering, Huaiyin Institute of Technology, Huaian 223003, China
3 Suzhou Changfeng Avionics Co., Ltd., Suzhou 215151, China
* Correspondence: taowang@nuaa.edu.cn

Abstract: To investigate the biomechanical performances of artificial cervical disc (ACD) prostheses, many studies have been conducted, either with cervical sections of cadavers under physiological loads or with block-like testing fixtures obeying the ASTM F2346 standard. Unfortunately, both methods are almost impossible to utilize for accurate results of lifetime anti-fatigue experiments for at least 10 million cycles due to the difficulties in cadaver preservation and great deviations of natural cervical bodies, respectively. Based on normal human cervical structural features, a novel specimen fixture was designed for testing the fatigue behavior of ACD prostheses under flexion, extension, and lateral bending conditions, with aspects of both structural and functional bionics. The equivalence between the biomimetic fatigue-testing fixture and the natural cervical sections was investigated by numerical simulations and mechanical experiments under various conditions. This study shows that this biomimetic fatigue-testing fixture could represent the biomechanical characteristics of the normal human cervical vertebrae conveniently and with acceptable accuracy.

Keywords: artificial cervical disc (ACD); specimen fixture; fatigue behavior; biomimetic

Citation: Cheng, X.; Bai, J.; Wang, T. Biomimetic Design of Fatigue-Testing Fixture for Artificial Cervical Disc Prostheses. *Metals* **2023**, *13*, 299. https://doi.org/10.3390/met13020299

Academic Editor: Alberto Campagnolo

Received: 27 December 2022
Revised: 27 January 2023
Accepted: 30 January 2023
Published: 1 February 2023

1. Introduction

Disc arthroplasty is a new surgical treatment for intervertebral degeneration and instability. Compared with traditional cervical fusion surgery, its advantages are that it restores the range of motion of the cervical spine and can lower the incidence of adjacent segment degeneration in the long term [1–4]. ACD prostheses are intended to bear alternating loads within the scope of physiology and should theoretically last for several decades in the body without failures. The life of ACD prostheses and their minimum acceptable clinical life are disputed, however. Titanium and titanium alloys are widely used in orthopedic hard tissue repair and artificial cervical intervertebral disc manufacture because of their good biocompatibility and non-toxicity [5,6]. The optimum life span has generally been demonstrated to be 80 million movements, while 10 million movements is suggested to be the ideal minimum testing cycle [7,8].

Traditional implant trials often choose a series of cervical spine specimens from cadaver donors moisturized with saline solution. Cervical spine specimens are dissected free from soft tissues, musculature, and single-segmental cervical intervertebral discs, while the ligaments and integrity of post-zygapophysial joints are carefully preserved, and then ACD prostheses are implanted for testing [9–12]. These specimen tests must be completed in a short time to avoid causing side effects in the process of the biological disintegration of cadavers [13–17]. Unfortunately, it is almost impossible to carry out a lifetime anti-fatigue experiment for at least 10 million cycles with fresh-frozen cadavers due to the time and cost limitations of cadaver preservation.

In dealing with such problems, the tests conducted by ASTM F2346 allow for the analysis of individual disc replacement devices and the comparison of the mechanical performance of multiple artificial intervertebral disc designs in a standard model [18–23]. Specialized test fixtures have been developed by leading machine manufacturers to conduct both static and cyclic testing of ACDs following the ASTM F2346 standard in recent decades. Due to the obvious difference between these test fixtures and the natural cervical spine, the obtained data about the static and dynamic behaviors of ACD prostheses are less accurate. Therefore, the results from these tests may not directly predict in vivo performance; however, they can be used to compare the mechanical performance of different ACD prostheses [19,20].

In dealing with the ultra-high cost and inevitable data deviation of the above-mentioned testing methods, the biomimetic methodology could be a promising solution for testing ACD prostheses with better accuracy and efficiency.

The objective of this study was to design a biomimetic fatigue-testing fixture using synthetic materials similar to human cervical vertebrae for cost-effective, accurate static and dynamic tests of ACDs, to evaluate the equivalence of the stresses and deformations (i.e., range of motion) of ACDs within the designed fatigue-testing fixture, and within C5–C6 cervical spinal segments. Fatigue simulations of ACDs within C5–C6 cervical spinal segments and within the fatigue-testing fixture and fatigue experiments of ACDs within the fatigue-testing fixture were also carried out. By comparing results, the feasibility of the designed biomimetic fatigue-testing fixture can be discussed thoroughly.

2. Materials and Methods

2.1. Biomimetic Design of Fatigue-Testing Fixture

2.1.1. Structural Bionics of Fatigue-Testing Fixture

The design of the biomimetic fatigue-testing fixture is shown in Figure 1. Based on normal human cervical structural features, four epoxy blocks filled with 70 wt% hydroxyapatite powder, the same content as natural bone, were used to simulate human cervical vertebrae, as the elastic modulus of epoxy blocks is close to that of human cervical vertebrae. A metallic flexible U-plate of the fatigue-testing fixture limited the movement of blocks to simulate the function of normal human cervical ligaments and facet joints.

In terms of the size measurement and deformation properties of the C5–C6 cervical spinal segments, as shown in Figure 1, the fixture is cost-effective and reasonable, composed of a cuboid block (01), three cylindrical blocks (02–04) and a U-plate (05). Among blocks 01–04, the cylindrical blocks 02–04 are concentric, whereas the position of cuboid block 01, as the site for compressive force to be applied, can be varied to form different loading conditions. According to preliminary estimations, the thickness and width of the U-plate (05) were 1~2 mm and 30~45 mm. The length, width, and height of the cuboid block (01) were 25~35 mm, 8~15 mm, and 8~15 mm. The radius and height of the cylindrical blocks (02–04) were 10~15 mm and 8~15 mm. The distance between the center of the cylindrical blocks (02–04) and the rear end of the U-plate (05) was 45~65 mm. Using numerical simulation analysis, the fixture design was further optimized. Additionally, the lower surface of the cylindrical block (02) and the upper surface of the cylindrical block (03) were polished to meet the body's normal physiological curvature of the cervical lordosis.

The fixture was able to work with the integration prostheses and the majority of other ACD prostheses, such as Dynamic Cervical Implant (DCI), Z-braze Dynamic Fusion, and Prestige LP, as shown in Figure 1. Among various ACDs, DCI is a single-component U-shaped implant maintaining the spinal kinematics, which imposes minimum influence on the adjacent soft tissues compared with other types of prostheses [24]. The U-shaped open structure of the DCI is more favorable for the direct observation of static and dynamic behaviors during the numerical simulations of implants. Therefore, DCI (14 mm long, 16 mm wide, and 0.8 mm thick) was selected as the ACD specimen for the study of the effectiveness of the designed biomimetic fatigue-testing fixture in substitution of the natural cervical spine in static and dynamic experiments.

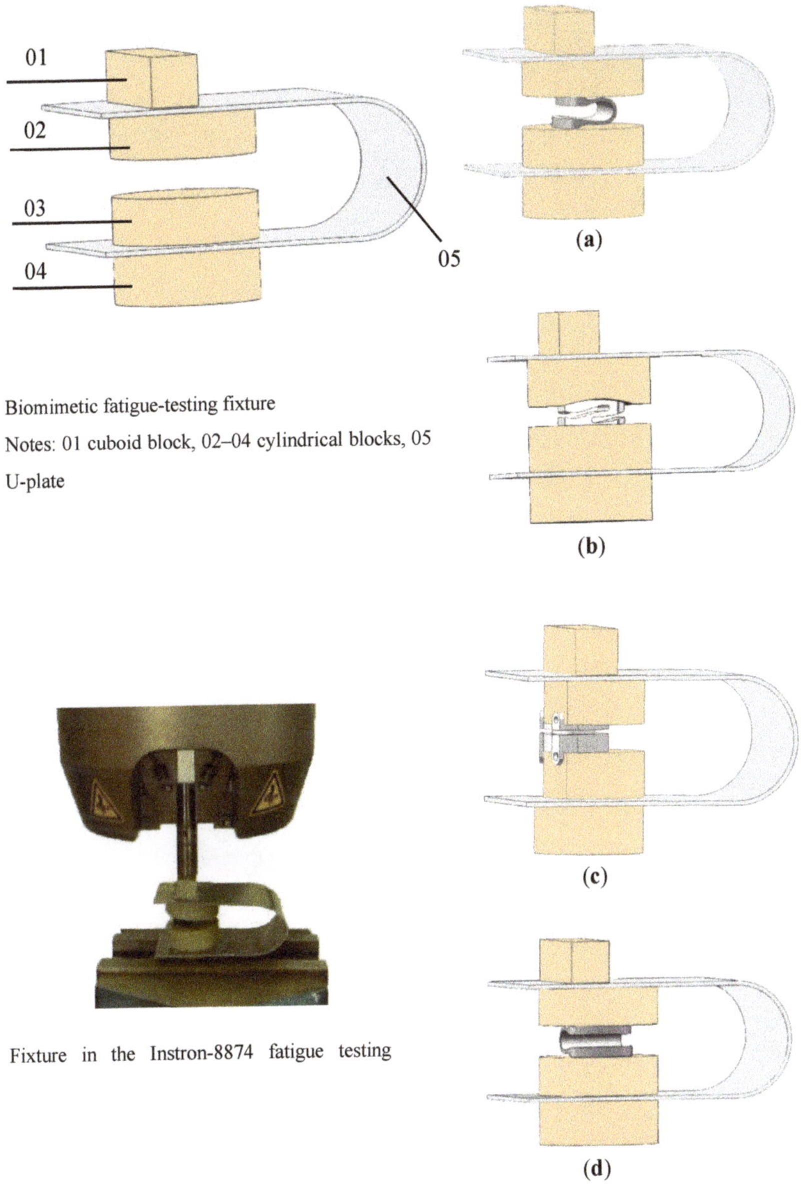

Figure 1. Assembly drawing of fixture and prostheses: (**a**) Dynamic Cervical Implants DCI under the flexion or extension condition; (**b**) Z-braze Dynamic Fusion under the flexion or extension condition; (**c**) Prestige LP under the flexion or extension condition; (**d**) DCI under the lateral bending condition.

2.1.2. Functional Bionics of Fatigue-Testing Fixture

The geometric models of the fatigue-testing fixture with DCI and the cervical bodies with DCI were established, which were input into the finite element analysis software ANSYS Workbench 16.0 (Ansys, Canonsburg, PA, USA) and assigned with corresponding material properties, as shown in Table 1 [25,26].

Table 1. Material properties of fatigue-testing fixture and spinal components used in the present model.

Description	Elastic Modulus (MPa)	Poisson's Ratio
Blocks	20,000	0.35
6061 Al alloy	70,000	0.3
Pure Ti	110,000	0.3
Ti6Al4V	110,000	0.3
Epoxy AB glue	3000	0.38
Cortical bone	10,000	0.4
Cancellous bone	100	0.29
Ligamentum flavum	1.5	0.3
Interspinous ligament	1.5	0.3
Capsular ligament	20	0.3

In the numerical calculation of the equivalent stress and deformation of DCI within the C5–C6 cervical spinal segments during the static test, the maximum routine loading parameters in biomechanical tests with cervical spines from cadaver donors were followed; namely, a 73.6 N preload was applied to the top surface of C5, with an extra 1.8 Nm flexion moment for flexion movement and a 1.8 Nm extension moment for extension or a 1.0 Nm lateral bending moment for bending, respectively, while the bottom surface of C6 was fixed in six degrees of freedom in the finite element model [26–28].

As for the simulations of the static test with the biomimetic fatigue-testing fixture, a 200 N load, the routine loading force in static and dynamic tests following ASTM F2346, was applied on the upper surface of the cuboid block 01, while the lower surface of the cylinder 04 was fixed in six degrees of freedom [19,29,30]. Furthermore, by finely adjusting the distance between the centers of DCI and the cuboid block 01, an extra equivalent moment can be obtained by multiplying the force on the top surface of the cuboid block 01 and the eccentric distance. Finally, the equivalent stress and deformation of DCI within the fatigue-testing fixture were calculated by using finite element simulation software ANSYS Workbench 16.0 (Ansys, Canonsburg, PA, USA). The equivalent stress and deformation of DCI within the fatigue-testing fixture can be made similar to those within human C5–C6 cervical spinal segments through further optimization of the parameters of the biomimetic fatigue-testing fixture.

2.2. *The Process of Fatigue Simulation and Fatigue Test*

The previous analysis results of DCI within C5–C6 cervical spinal segments and within the optimized fatigue-testing fixture were input into the FE-SAFE6.5 software (Dassault Systèmes Simulia Corp., Providence, RI, USA). According to the estimation methods of Seegers' material data in FE-SAFE6.5 software, SN curves were generated by inputting the tensile strength and elastic modulus, which were modified afterward [31,32]. A frequency of 10 Hz was applied with a triangular-wave load channel file. The Goodman method was used for mean stress correction, and the fatigue full-life analysis was carried out according to SN curves. After the fatigue calculation, the analysis results were re-input into ANSYS Workbench 16.0 (Ansys, Canonsburg, PA, USA) for post-processing.

In the present fatigue experiments, the DCI was fixed between the cylindrical hydroxyapatite-filled epoxy blocks 02 and 03 by epoxy AB glue in the biomimetic fatigue-testing fixture of the optimized parameters. It is used to simulate the bony fusion between the upper and lower surfaces of the DCI prosthesis and the corresponding vertebral body after the implant. The loading parameters of the cervical segment biomechanical tests were consistent with static and dynamic tests based on ASTMF2346 by adjusting the distance between the geometric center of the block and the rotating center of the artificial cervical disc. Then, the fixture, together with the DCI, was clamped between the vise and actuator of an Instron-8874 fatigue testing machine (Instron Corporation, Canton, MA, USA), as shown in Figure 1. During fatigue experiments, the fixture was loaded with the corresponding loadings at the calculated eccentric position to provide different moments. Finally, fatigue tests were carried out until fatigue failure occurred; if not, the tests were continued until 80 million cycles were reached on the biomimetic fatigue-testing fixture.

3. Results

3.1. Optimization of the Biomimetic Fatigue-Testing Fixture

The maximum deformation of pure Ti DCI within human C5–C6 cervical segments under the flexion condition was numerically calculated as 0.57 mm, as shown in Figure 2. On the foundation that the stress of the DCI within the fatigue-testing fixture is similar to that within C5–C6 cervical spinal segments, the deformations of pure Ti DCI within the designed fatigue-testing fixture of different parameters were numerically calculated and shown in Figures 3 and 4. With the other factors being equal, the maximum deformation decreased markedly with an increase in the elastic modulus of the material and the thickness and width of the U-plate 05, respectively, as shown in Figure 3, whereas the length, width, and height of the cuboid block 01, as well as the radius and height of the cylindrical blocks 02–04, lacked an obvious influence on the DCI's maximum deformation, as shown in Figure 4a–e. Additionally, the DCI's maximum deformation increased slowly as the distance between the center of the cylindrical blocks 02–04 and the rear end of the U-plate 05 increased, as shown in Figure 4f. Likewise, the influencing tendencies of the above-mentioned factors on the deformation of DCI under the flexion condition coincide with those of DCI under either extension or lateral bending conditions. Finally, the optimizations of the various designing factors were conducted by keeping the coincidence of the stresses and deformations of the DCI within the designed biomimetic fatigue-testing fixture and within human C5–C6 cervical segments.

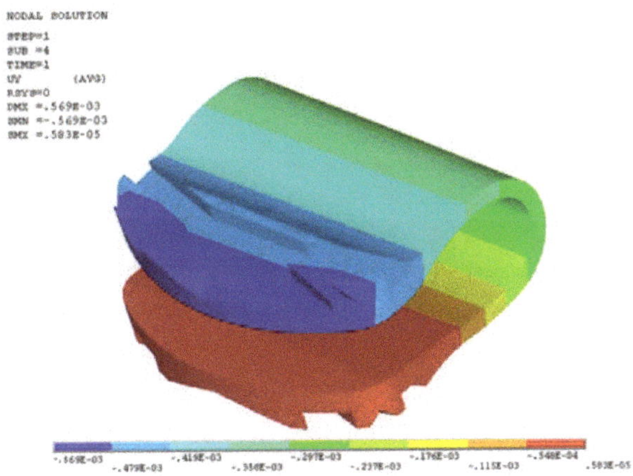

Figure 2. The deformation of pure Ti DCI within human C5–C6 cervical segments under the flexion condition.

3.2. Simulation of DCI within the Optimized Fixture under Static Mode

The contours of the equivalent stress of pure Ti DCI within the C5–C6 cervical spinal segments and within the optimized fatigue-testing fixture under the flexion condition are shown in Figure 5. The maximum equivalent stress of pure Ti DCI within the fatigue-testing fixture was 396.5 MPa, which agreed well with 394.6 MPa, the result of DCI within the C5–C6 cervical segments. More importantly, both maximum equivalent stresses appeared in the same location of the DCI. Furthermore, the contours of the equivalent stress of the DCIs of pure Ti and Ti6Al4V were simulated under various loading conditions within the fatigue-testing fixture, which were similar to those of the DCIs within the C5–C6 cervical spinal segments, as plotted in Figure 6.

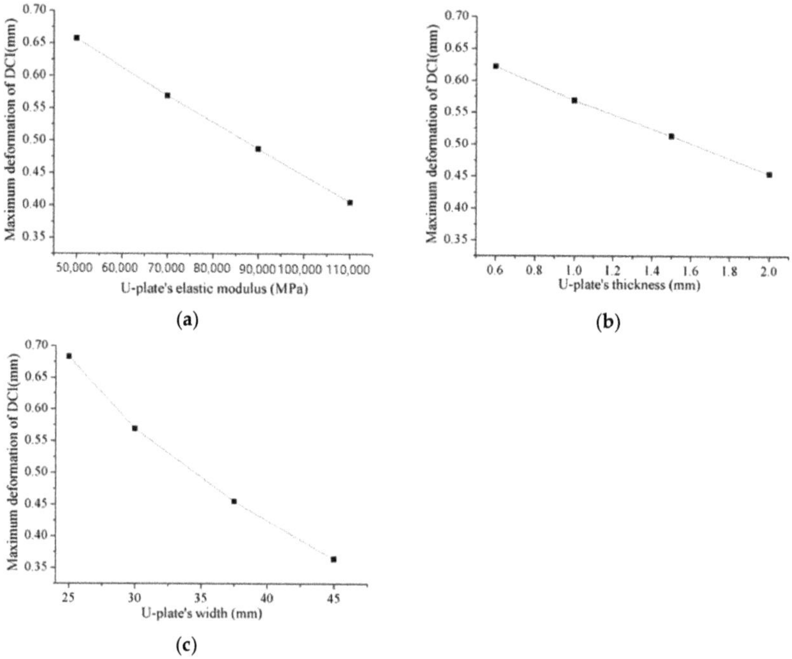

Figure 3. Influence of the U-plate's elastic modulus and size on the DCI's maximum deformation: (**a**) U-plate thickness and width were fixed at 1 mm and 30 mm; (**b**) elastic modulus and width at 70,000 MPa and 30 mm; (**c**) elastic modulus and thickness at 70,000 MPa and 1 mm.

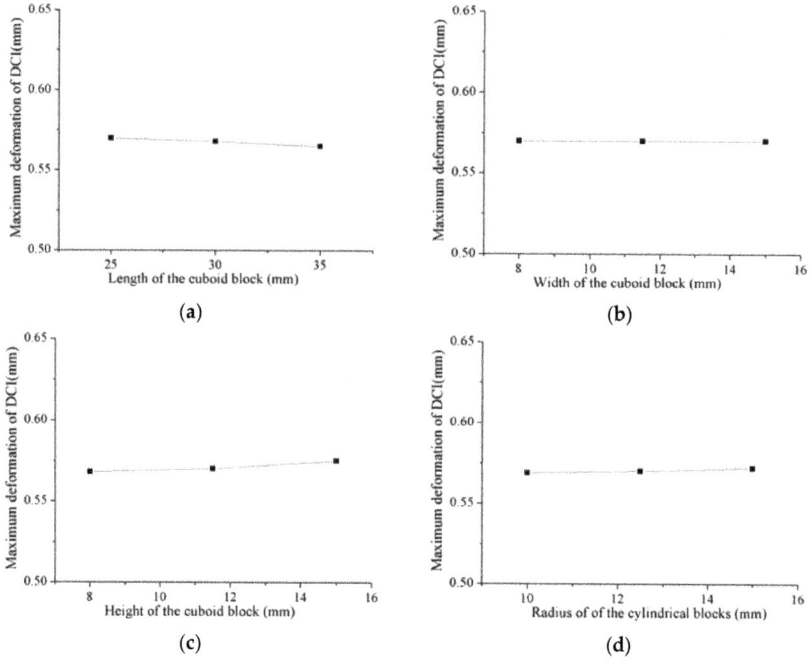

Figure 4. *Cont.*

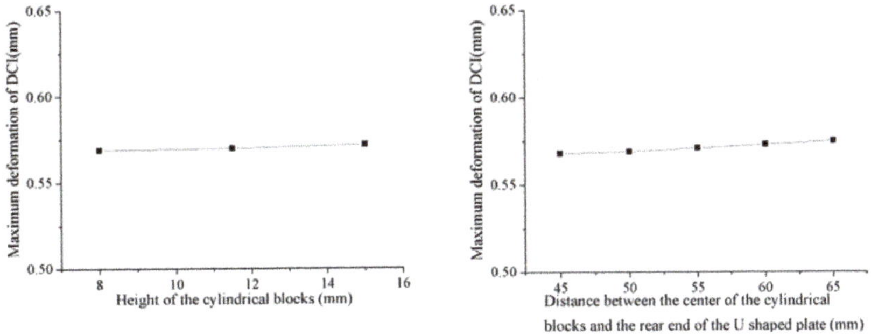

(e) (f)

Figure 4. Influence of the geometric sizes and position of the blocks on the DCI's maximum deformation: (**a**) the length of the cuboid block; (**b**) the width of the cuboid block; (**c**) the height of the cuboid block; (**d**) the radius of the cylindrical blocks; (**e**) the height of the cylindrical blocks; (**f**) the distance between the center of the cylindrical blocks and the rear end of the U-plate, assuming other factors being equal.

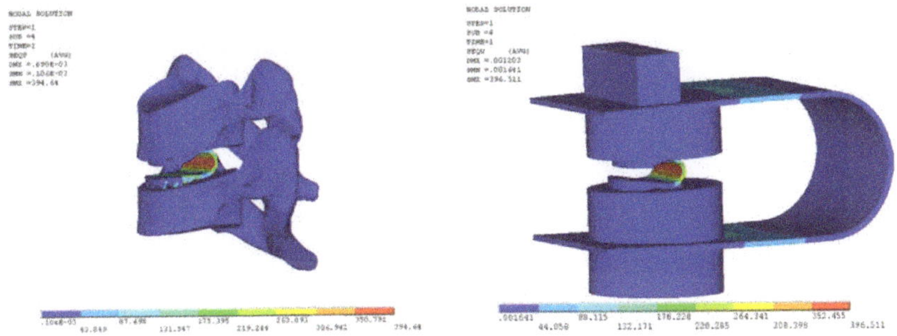

(a) (b)

Figure 5. The contours of the equivalent stress of pure Ti DCI under the flexion condition: (**a**) within C5–C6 cervical spinal segments; (**b**) within the fatigue-testing fixture.

3.3. Fatigue Simulation and Fatigue Experiment

Figure 7 shows the contours of the fatigue life of pure Ti DCI within the C5–C6 cervical spinal segments and within the optimized fatigue-testing fixture under the flexion condition. The minimums of the simulated fatigue life of pure Ti DCI within the C5–C6 cervical spinal segments and within the fatigue-testing fixture were 22.397 million cycles ($N = 10^{7.3502} = 22{,}397{,}000$) and 21.478 million cycles ($N = 10^{7.332} = 21{,}478{,}000$), respectively. The fatigue results of the DCIs of pure Ti and Ti6Al4V within the C5–C6 cervical spinal segments and within the fatigue-testing fixture were simulated under various loading conditions, as summarized in Table 2. The simulated fatigue life of titanium alloy exceeds 80 million cycles during flexion. In the process of extension and lateral bending, the simulated fatigue life of both pure titanium and titanium alloy exceeds 80 million times.

The fatigue-testing results of the DCIs of pure Ti and Ti6Al4V within the fatigue-testing fixture were obtained under various loading conditions by using an Instron-8874 fatigue-testing machine, as plotted in Table 2. It was shown that the experimental fatigue life of pure Ti DCI within the fatigue-testing fixture under the flexion condition was 35.645 million cycles, whereas the fatigue lives of pure Ti DCIs under other experimental conditions, as well as Ti6Al4V DCIs under all experimental conditions, were more than 80 million cycles.

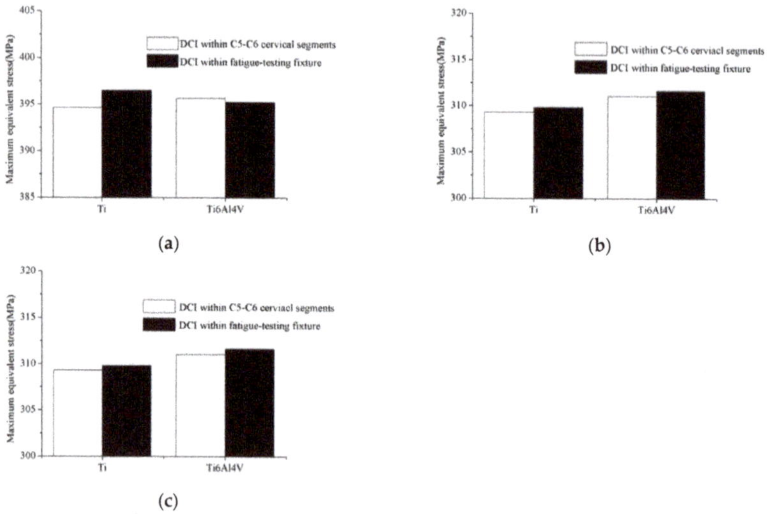

Figure 6. The maximum equivalent stress of DCI within C5–C6 cervical segments and within the fatigue-testing fixture in (**a**) flexion, (**b**) extension, and (**c**) lateral bending movements.

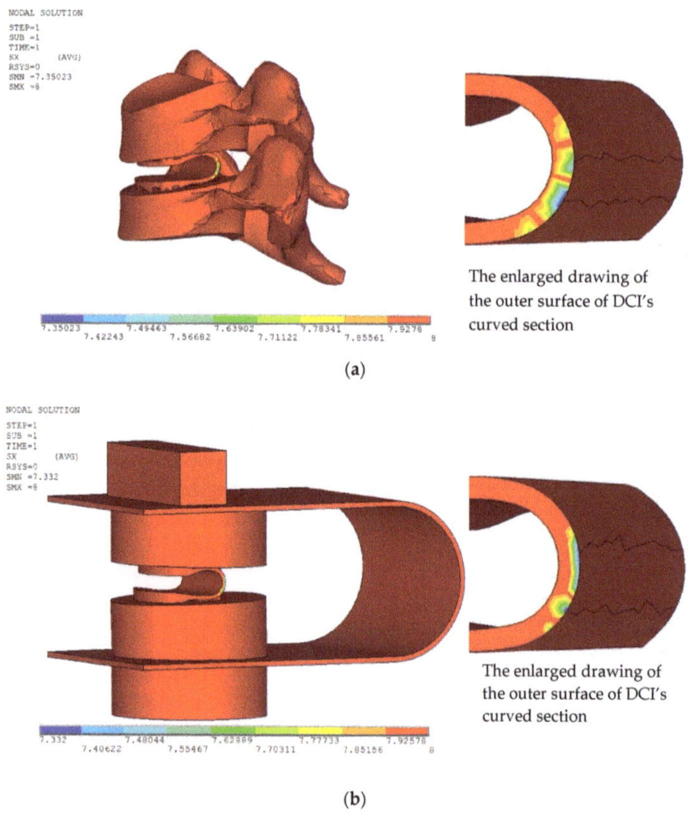

Figure 7. The contours of the fatigue life of pure Ti DCI under the flexion condition: (**a**) within C5–C6 cervical spinal segments; (**b**) within the fatigue-testing fixture.

Table 2. Simulated fatigue results and fatigue-testing results.

Type of Load	DCI Material	Simulated Fatigue Life of DCI within C5–C6 Cervical Spinal Segments (Million Cycles)	Simulated Fatigue Life of DCI within Fatigue-Testing Fixture (Million Cycles)	Fatigue-Testing Life of DCI within Fatigue-Testing Fixture (Million Cycles)
Flexion	Ti	22.397	21.478	35.645
	Ti6Al4V	≥80	≥80	≥80
Extension	Ti	≥80	≥80	≥80
	Ti6Al4V	≥80	≥80	≥80
Lateral bending	Ti	≥80	≥80	≥80
	Ti6Al4V	≥80	≥80	≥80

4. Discussion

Human physiological motion is complicated, comprehensive, and cooperative, which is difficult to represent accurately. However, its main functions are highlighted by mimicking the main biological structures and control principles of human cervical vertebrae.

4.1. The Rationality of the Static Load Settings

Notably, 200 N is also the routine maximum fatigue compressive force applied in the dynamic tests of ACDs according to ASTM F2346 [21,29,30]. In order to bridge the loading parameters in biomechanical tests with cervical spines from cadaver donors and in static and dynamic tests following ASTM F2346, an extra equivalent moment can be obtained by finely adjusting the eccentric distance between the centers of the cuboid block 01 and ACD during the finite element simulations of the static and fatigue experiments. For example, in the flexion movement, a 1.8 Nm flexion moment and 73.6 N preload with a 6 mm eccentric distance between the center of the implant position of ACD and the center of C5–C6 cervical spinal segments were applied onto the top surface of C5; therefore, the comprehensive loading moment was 73.6 N × 6 mm + 1.8 Nm = 2.242 Nm. According to ASTM F2346 test methods, a similar comprehensive load in the fatigue-testing fixture can be achieved (i.e., 200 N × 11.2 mm = 2.240 Nm) only by adjusting the eccentric distance between the force loading position (cuboid block 01) and the center of ACD to 11.2 mm [29,33]. Likewise, identical comprehensive loads can be obtained for either extension or lateral bending movements.

Meanwhile, the extra equivalent moment was also stable in the process of motion due to the tiny deformation of ACDs in the fatigue test. In consideration of experimental conditions and the various testing requirements of ACDs, the above loading methodology is not merely reasonable but also easily achieved.

4.2. Optimization of the Biomimetic Fatigue-Testing Fixture

When the elastic modulus of U-plate 05 reached 70,000 MPa, the deformation of the DCI was 0.57 mm, which is the same value as that of the DCI with human C5–C6 cervical spinal segments, as shown in Figure 3a. Among a variety of candidate materials, 6061 Al alloy was the most suitable. Simultaneously, the optimized thickness and width of the U-plate 05 were confirmed as 1 mm and 30 mm, respectively, according to Figure 3b,c. The geometric sizes of the cuboid block 01 and the cylindrical blocks 02–04 had no obvious influence on the DCI's maximum deformation, as shown in Figure 4. Therefore, the sizes of the blocks were determined according to those of the cervical vertebrae, while the distance between the center of the cylindrical blocks 02–04 and the rear end of the U-plate 05 should be inclined to that between the intervertebral disc and ligaments.

The optimized parameters of the biomimetic fatigue-testing fixture are as follows: 6061 Al alloy is suitable for the U-plate 05; the thickness and width of the U-plate 05 are 1 mm and 30 mm; the hydroxyapatite-filled epoxy block 01 is a 25 mm long, 10 mm wide and 10 mm high cuboid; the radius and height of the hydroxyapatite-filled epoxy cylindrical blocks 02–04 are 12 mm and 10 mm; and the distance between the center of the cylindrical blocks 02–04 and the rear end of the U-plate 05 is 50~55 mm.

By employing these loading conditions and the optimized fixture, the simulated results of the maximum equivalent stress of the DCI within the C5–C6 cervical segments and within the fatigue-testing fixture present a series of consistencies in flexion, extension, and lateral bending movements, as shown in Figure 6.

4.3. The Safety of the Biomimetic Fatigue-Testing Fixture

The safety of the fixture is its foundation and premise during long periods of cyclic loading. The simulated fatigue lives of the U-plate and the blocks were more than 80 million cycles, as shown in Figure 7. Furthermore, the failure of the biomimetic fatigue-testing fixture did not occur in the 80-million-cycle fatigue experiments. Therefore, it can be concluded that the fixture is highly safe.

4.4. The Equivalence between the Biomimetic Fatigue-Testing Fixture and the Natural Cervical Sections

As shown in Figure 5, the curved section of the DCI is prone to forming crack sources due to large stresses; the cracks could propagate gradually during long periods of cyclic loading and finally cause the fatigue fracture of the DCI when the cyclic times accumulate beyond its fatigue life.

Under the circumstance that the equivalent stress and deformation of the DCI within C5–C6 cervical segments are almost the same as those within the biomimetic fatigue-testing fixture, the results of both fatigue simulations coincide well with the experimental results. The calculated fatigue lives of DCI within the C5–C6 cervical spinal segments and within the fatigue-testing fixture were 22.397 million cycles and 21.478 million cycles, respectively, which agree well with the experimental fatigue life of 35.645 million cycles, as shown in Table 2. It is noticeable that the simulated fatigue lives and possible sites for the fatigue failure of pure Ti DCI were almost same whether it was fixed within C5–C6 cervical spinal segments or within the fatigue-testing fixture, as shown in Figure 7.

In brief, the prostheses within the fatigue-testing fixture under the loads according to ASTM F2346 can achieve a functionally equivalent result to that under the biomechanical loads within normal cervical vertebrae.

4.5. Limitations of the Biomimetic Fatigue-Testing Fixture

The present study of biomimetic fatigue-testing fixtures has two limitations. Firstly, actual cervical movement includes not only flexion, extension, lateral bending, and axial torsion but also the combinations of single movement patterns within the scope of physiology. The biomechanical axial torsion and preload cannot be equivalent to the moment, which relates to the compressive force perpendicular to the surface of cuboid block 01, because they are not in a common plane. Therefore, the biomimetic fixture cannot meet the requirements for the torsion condition. Additionally, the biomimetic fatigue-testing fixture is only suitable for single load patterns, such as flexion, extension, and lateral bending. Unfortunately, single load patterns may be considered to be clinically unrealistic.

Secondly, muscle forces in spinal motions should not be neglected. Muscles in the loading spine generate spinal reaction forces, which can occupy the main portion of the total axial compression and shear forces on the spine, further affecting the life of ACD prostheses [34]. Simultaneously, posterior muscles can assist with balance in flexion postures; accordingly, anterior muscles act in the same role in extension postures. They can reduce the reaction forces of the lower joints and keep the spine steady [35]. These aspects are significant for ACD prostheses, especially when dynamic or impact loading is applied [36,37]. Unfortunately, synergism among the muscles is difficult to investigate because specimens in vitro cannot mimic the role of muscles well [34].

5. Conclusions

In summary, a novel specimen fixture has been designed for testing the fatigue behavior of ACD prostheses with aspects of both structural and functional bionics. The equivalence between the designed biomimetic fixture and the natural cervical sections has been verified by numerical simulations and mechanical experiments. This biomimetic fatigue-testing fixture represented the biomechanical characteristics of normal human cervical vertebrae with considerable accuracy. The novel specimen fixture provides a convenient and accurate way to research and evaluate the fatigue behavior of ACD prostheses.

Author Contributions: X.C., J.B. and T.W. contributed substantially to the conception and design of the experiments. X.C. and J.B. conducted experiments and wrote the manuscript. T.W. conducted data analyses. All authors have read and agreed to the published version of the manuscript.

Funding: This research was funded by the International Science and Technology Cooperation Programme (2018YFE0194100), the National Natural Science Foundation of China (51875231), and the Priority Academic Program Development of Jiangsu Higher Education Institutions.

Institutional Review Board Statement: Not applicable.

Informed Consent Statement: Not applicable.

Data Availability Statement: The original contributions presented in the study are included in the article; further inquiries can be directed to the corresponding author.

Conflicts of Interest: The authors declare no conflict of interest.

References

1. Hilibrand, A.S.; Robbins, M. Adjacent segment degeneration and adjacent segment disease: The consequences of spinal fusion? *Spine J.* **2004**, *4*, S190–S194. [CrossRef] [PubMed]
2. Matsumoto, M.; Okada, E.; Ichihara, D.; Watanabe, K.; Chiba, K.; Toyama, Y. Adjacent segment disease and degeneration after anterior cervical decompression and fusion. *Neurosurg. Q.* **2010**, *20*, 15–22. [CrossRef]
3. Fiani, B.; Nanney, J.M.; Villait, A.; Sekhon, M.; Doan, T. Investigational research: Timeline, trials, and future directions of spinal disc arthroplasty. *Cureus* **2021**, *13*, 16739. [CrossRef] [PubMed]
4. Joaquim, A.F.; Makhni, M.C.; Riew, K.D. Evidence-based use of arthroplasty in cervical degenerative disc disease. *Int. Orthop.* **2019**, *43*, 767–775. [CrossRef] [PubMed]
5. Lee, H.; Lee, M.; Han, G.; Kim, H.; Song, J.; Na, Y.; Yoon, C.; Oh, S.; Jang, T.; Jung, H. Customizable design of multiple-biomolecule delivery platform for enhanced osteogenic responses via 'tailored assembly system'. *Bio-Des. Manuf.* **2022**, *5*, 451–464. [CrossRef]
6. Lee, H.; Lee, M.; Cheon, K.; Kang, I.; Park, C.; Jang, T.; Han, G.; Kim, H.; Song, J.; Jung, H. Functionally assembled metal platform as lego-like module system for enhanced mechanical tunability and biomolecules delivery. *Mater. Design.* **2021**, *207*, 109840. [CrossRef]
7. Gloria, A.; Causa, F.; De Santis, R.; Netti, P.A.; Ambrosio, L. Dynamic-mechanical properties of a novel composite intervertebral disc prosthesis. *Mater. Med.* **2007**, *18*, 2159–2165. [CrossRef]
8. Rosa, G.L.; Clienti, C.; Corallo, D. Design of a new intervertebral disc prosthesis. *Mater. Today Proc.* **2019**, *7*, 529–536. [CrossRef]
9. Barker, J.B.; Cronin, D.S.; Chandrashekar, N. High rotation rate behavior of cervical spine segments in flexion and extension. *J. Biomech. Eng.* **2014**, *136*, 121004. [CrossRef]
10. Nuckley, D.J.; Linders, D.R.; Ching, R.P. Developmental biomechanics of the human cervical spine. *J. Biomech.* **2013**, *46*, 1147–1154. [CrossRef]
11. Lou, J.; Hao, L.; Li, Y.; Orthopedics, D.O.; Hospital, W.C.; University, S. Biomechanical evaluation of one new artificial cervical disc prosthesis. *Orthop. Biomech. Mater. Clin. Study* **2016**, *13*, 10–13.
12. Patwardhan, A.G.; Havey, R.M. Prosthesis design influences segmental contribution to total cervical motion after cervical disc arthroplasty. *Eur. Spine J.* **2020**, *29*, 2713–2721. [CrossRef]
13. Toen, C.V.; Melnyk, A.D.; Street, J.; Oxland, T.R.; Cripton, P.A. The Effects of Lateral Eccentricity on Failure Loads and Injuries of the Cervical Spine in Head-First Impacts. In Proceedings of the Ohio State University's 10th Annual Injury Biomechanics Symposium, Columbus, OH, USA, 18–20 May 2014; p. 18.
14. Rezaei, A.; Giambini, H.; Carlson, K.D.; Xu, H.; Lu, L. Mechanical testing setups affect spine segment fracture outcomes. *J. Mech. Behav. Biomed.* **2019**, *100*, 103399. [CrossRef]
15. Rahm, M.; Brooks, D.; Harris, J.; Hart, R.; Hughes, J.; Ferrick, B.; Bucklen, B. Stabilizing effect of the rib cage on adjacent segment motion following thoracolumbar posterior fixation of the human thoracic cadaveric spine: A biomechanical study. *Clin. Biomech.* **2019**, *70*, 217–222. [CrossRef]

16. Mannen, E.M.; Friis, E.A.; Sis, H.L.; Wong, B.M.; Cadel, E.S.; Anderson, D.E. The rib cage stiffens the thoracic spine in a cadaveric model with body weight load under dynamic moments. *J. Mech. Behav. Biomed.* **2018**, *84*, 258. [CrossRef] [PubMed]
17. Shen, F.H.; Woods, D.; Miller, M.; Murrell, B.; Vadapalli, S. Use of the dual construct lowers rod strains in flexion-extension and lateral bending compared to two-rod and two-rod satellite constructs in a cadaveric spine corpectomy model. *Spine J.* **2021**, *21*, 2104–2111. [CrossRef]
18. Holsgrove, T.P.; Miles, A.W.; Gheduzzi, S. The application of physiological loading using a dynamic, multi-axis spine simulator. *Med. Eng. Phys.* **2017**, *41*, 74–80. [CrossRef]
19. Phillips, F.M.; Geisler, F.H.; Gilder, K.M.; Reah, C.; Howell, K.M.; McAfee, P.C. Long-term outcomes of the us fda ide prospective, randomized controlled clinical trial comparing pcm cervical disc arthroplasty with anterior cervical discectomy and fusion. *Spine* **2015**, *40*, 674–683. [CrossRef] [PubMed]
20. Graham, J.; Estes, B.T. What standards can (and can't) tell us about a spinal device. *SAS J.* **2009**, 3, 178–183. [CrossRef] [PubMed]
21. Mannen, E.M. Mechanical Testing of the Thoracic Spine and Related Implants. In *Mechanical Testing of Orthopaedic Implants*; Woodhead Publishing: Sawston, UK, 2017; pp. 143–160.
22. Cza, B.; Emm, C.; Hls, D.; Esc, D.; Bmw, D.; Ww, B.; Bo, C.B.; Eaf, D.; Deaa, E. Moment-rotation behavior of intervertebral joints in flexion-extension, lateral bending, and axial rotation at all levels of the human spine: A structured review and meta-regression analysis. *J. Biomech.* **2020**, *100*, 109579.
23. Sherrill, J.T.; Siddicky, S.F.; Davis, W.D.; Chen, C.; Mannen, E.M. Validation of a custom spine biomechanics simulator: A case for standardization. *J. Biomech.* **2019**, *98*, 109470. [CrossRef] [PubMed]
24. Mo, Z.J.; Zhao, Y.B.; Wang, L.Z.; Sun, Y.; Zhang, M.; Fan, Y.B. Biomechanical effects of cervical arthroplasty with u-shaped disc implant on segmental range of motion and loading of surrounding soft tissue. *Eur. Spine J.* **2014**, *23*, 613–621. [CrossRef] [PubMed]
25. Teo, E.C.; Ng, H.W. Evaluation of the role of ligaments, facets and disc nucleus in lower cervical spine under compression and sagittal moments using finite element method. *Med. Eng. Phys.* **2001**, *23*, 155–164. [CrossRef] [PubMed]
26. Vette, A.H.; Yoshida, T.; Thrasher, T.A.; Masani, K.; Popovic, M.R. A comprehensive three-dimensional dynamic model of the human head and trunk for estimating lumbar and cervical joint torques and forces from upper body kinematics. *Med. Eng. Phys.* **2012**, *34*, 640–649. [CrossRef]
27. Nimbarte, A.D.; Zreiqat, M.; Ning, X. Impact of shoulder position and fatigue on the flexion-relaxation response in cervical spine. *Clin. Biomech.* **2014**, *29*, 277–282. [CrossRef]
28. Cheng, X.; Wang, T.; Pan, C. Finite element analysis and validation of segments c2-c7 of the cervical spine. *Metals* **2022**, *12*, 2056. [CrossRef]
29. Kim, S.B.; Bak, K.H.; Cheong, J.H.; Kim, J.M.; Kim, C.H.; Oh, S.H. Biomechanical testing of anterior cervical spine implants: Evaluation of changes in strength characteristics and metal fatigue resulting from minimal bending and cyclic loading. *J. Korean Neurosurg. Soc.* **2005**, *37*, 217–222.
30. Bai, C.; Wei, W.; Hou, D. Biomechanical analysis of a new titanium rubber cervical disc prosthesis. *Chin. J. Spine Spinal Cord.* **2014**, *24*, 752–756.
31. Elder, J.E.; Thamburaj, R.; Patnaik, P.C. Optimising ion implantation conditions for improving wear, fatigue, and fretting fatigue of ti-6ai-4v. *Surf. Eng.* **1989**, *5*, 55–79. [CrossRef]
32. Kim, W.J.; Hyun, C.Y.; Kim, H.K. Fatigue strength of ultrafine-grained pure ti after severe plastic deformation. *Scripta Mater.* **2006**, *54*, 1745–1750. [CrossRef]
33. Espinoza-Larios, A.; Ames, C.P.; Chamberlain, R.H.; Sonntag, V.; Dickman, C.A.; Crawford, N.R. Biomechanical comparison of two-level cervical locking posterior screw/rod and hook/rod techniques. *Spine J.* **2007**, *7*, 194–204. [CrossRef]
34. Arjmand, N.; Gagnon, D.; Plamondon, A.; Shirazi-Adl, A.; Larivière, C. Comparison of trunk muscle forces and spinal loads estimated by two biomechanical models. *Clin. Biomech.* **2009**, *24*, 533–541. [CrossRef] [PubMed]
35. Toosizadeh, N.; Haghpanahi, M. Generating a finite element model of the cervical spine: Estimating muscle forces and internal loads. *Sci. Iran.* **2011**, *18*, 1237–1245. [CrossRef]
36. Li, Y.; Lewis, G. Influence of surgical treatment for disc degeneration disease at c5–c6 on changes in some biomechanical parameters of the cervical spine. *Med. Eng. Phys.* **2010**, *32*, 595–603. [CrossRef] [PubMed]
37. Panjabi, M.M.; Cholewicki, J.; Nibu, K.; Grauer, J.N.; Babat, L.B.; Dvorak, J. Mechanism of whiplash injury. *Clin. Biomech.* **1998**, *13*, 239–249. [CrossRef]

metals

Article

Finite Element Analysis and Validation of Segments C2-C7 of the Cervical Spine

Xuejin Cheng [1], Tao Wang [1,*] and Changjiang Pan [2]

1 College of Material Science and Technology, Nanjing University of Aeronautics and Astronautics, Nanjing 211106, China
2 Faculty of Mechanical and Material Engineering, Jiangsu Provincial Engineering Research Center for Biomaterials and Advanced Medical Devices, Huaiyin Institute of Technology, Huai'an 223003, China
* Correspondence: taowang@nuaa.edu.cn

Abstract: As an important part of the human spine, the cervical spine has a complex structure and easily suffers from diseases. Analysis of the biomechanical mechanism of cervical spine structure using the finite element model is not only helpful for the diagnosis, treatment and prevention of cervical spine diseases but also has positive significance for the performance evaluation of cervical spine implants. In this paper, a method of establishing a cervical C2-C7 finite element model based on CT image data is studied. Through the preprocessing of cervical CT images, the C2-C7 three-dimensional finite element model of the cervical spine was established. The pure moment loads of 0.33 Nm, 0.5 Nm, 1 Nm, 1.5 Nm and 2 Nm were applied to simulate flexion/extension, and the moment of 1 Nm was used to simulate the left and right lateral bending and axial rotation of the cervical spine. The relative range of motion (ROM) between each vertebral body was calculated. At the same time, the stress on some segments under axial load was analyzed. The results were basically consistent with the experimental data of in vitro studies, which verified the validity of the model.

Keywords: cervical spine; biomechanics; finite element model (FEM); range of motion (ROM)

Citation: Cheng, X.; Wang, T.; Pan, C. Finite Element Analysis and Validation of Segments C2-C7 of the Cervical Spine. *Metals* **2022**, *12*, 2056. https://doi.org/10.3390/met12122056

Academic Editor: Dmytro Orlov

Received: 8 October 2022
Accepted: 25 November 2022
Published: 29 November 2022

1. Introduction

The spine consists of the cervical vertebra, the thoracic vertebra, the lumbar vertebra and the sacrum. With its more complicated structure and larger ranges of motion compared to the thoracic vertebra and lumbar vertebra, the cervical vertebra plays a significant role in people's daily lives. With the development of modern industry, the change in people's work and lifestyle as well as the population aging, the incidence of cervical spondylosis has shown a rising trend year by year, which has brought great inconvenience to people's daily lives. The prevention of cervical spondylosis and the improvement of treatment effectiveness have always been goals to strive for. The statistics of the motion tendency and the stress distribution of the cervical vertebra under different loads can be obtained through experiments in vivo and in vitro [1–5]. However, their applications are subject to certain restrictions due to the exceptionality of the human body, resulting in high cost, high risk, and specificity of experimental results. Belytschko et al. [6] and other researchers established the two-dimensional finite element model early in 1974 to evaluate the material attribute of the intervertebral disc and to predict its mechanical behavior under axial loads. However, due to the complex anatomical structure of the cervical vertebra, the finite element model developed slowly. Not until the 1990s did Bozic et al. [7] and Teo et al. [8], along with their teams, accomplish the three-dimensional reconstruction of the vertebral body using CT data and a three-dimensional digitizer. Based on this, the more lifelike and fine-mesh finite element model of the single vertebra was constructed, although there was still a great discrepancy between the model and the actual conditions, especially in terms of less consideration given to soft tissue and the single material property. Since then, with the development of computer software, hardware as well as the advancement of medical imaging technology, the finite element model of the cervical spine has also developed from an initially single-level model to a multi-level one and has included the entire cervical level [9–13]. The material model of intervertebral disc and ligament has also been further refined.

The FEM of the cervical spine can be divided into dynamic analysis and static analysis depending on the application scenario. With respect to dynamic analysis of the cervical spine, Cronin's project team [11,14,15] from Canada performed in-depth research covering 3D modeling, meshing, and material model constitution, and also analyzed the mechanism of cervical spine soft tissue injury in frequent traffic accidents. Mustafy et al. [16] introduced time and rate-dependent material laws in the finite element model of segments C2-C3 to simulate the mechanism of cervical ligament injury under complex loading. Meyer et al. [17] simulated the forces and moments of the cervical spine segments in cases of a frontal impact with an experimentally validated finite element model. As for static analysis, Zhang et al. [18] built a finite element model of the complete cervical spine from C0 to C7 and compared the model with experimental results in terms of flexion/extension, lateral bending and axial rotation under a load of 1.0 Nm. Limited by the conditions back then, the material model settings were relatively simple, producing larger differences from the experimental results in terms of lateral bending. Kallemeyn et al. [19] built a finite element model of segments C2-C7 based on multiblock meshing techniques and specific specimen data; the material properties of the model were then calibrated using the experimental results of the specimens. Erbulut et al. [20] also built a finite element model of C2-T1 based on the multiblock meshing technique to study the effect of soft tissue on cervical spine stability. To promote the exchange of scientific research on finite element models of the cervical spine, Herron et al. [21] built the finite element models of the cervical spine of females at different ages. They noticed that despite the same material properties and boundary conditions, their biomechanical properties differed markedly, and the authors considered that was because of differences in the morphology of the models.

With the improvement of the finite element model, its applications in medical engineering widen increasingly. By creating finite element models of different cervical spine surgeries, the influence of different surgical methods on range of motion and intervertebral disc stress can be evaluated [22–24]. Nishida [25] established three different models of cervical soft tissue injury and analyzed the risk factors of cervical instability. In addition, the cervical spine finite element model can also be used to analyze the biomechanics of implanted medical devices and evaluate their performance [26,27]. In the field of traditional Chinese medicine, some researchers have also explored the application of finite element model in the treatment of cervical spondylosis with traditional traction and other conservative methods [28]. With the increasingly wide applications of the finite element method in spine research, it has become urgent to find a quick and convenient establishment of the finite element model for the spine. Targeting the cervical segments C2-C7, this paper is meant to provide a feasible solution for the finite element analysis of the complex spine structure based on the strong mesh processing function of the general pre-processing software and the analysis function of the professional finite element software.

2. Materials and Methods

2.1. Finite Element Model Establishment

The experimental data were obtained from the medical image database. The CT layer thickness was 0.625 mm, and the image resolution was 512 × 512 pixels, including the entire cervical spine. Firstly, the CT data were imported into medical image processing software, and the soft tissue images around the bone were removed using layer-by-layer segmentation, and the CT images of each tomography were edited and processed to complete the three-dimensional reconstruction of the cervical vertebra bone model. The model was imported into the reverse engineering software Geomagic 2012 (3D Systems Corporation, Rock Hill, SC, USA) in STL format, and the NURBS surface of the vertebral body was fitted by its functions, such as accurate surface, which was saved as an IGES file. The IGES file of the vertebral body was imported into the general finite element preprocessing software Hypermesh 13.0 (Altair Engineering Inc, Troy, MI, USA) to complete the partition of the vertebral body mesh. Based on the physiological and anatomical structure of cervical spine vertebra, the vertebral body is divided into the cortical bone, cancellous bone, endplate and posterior structure of the vertebral body. The correct choice of element for a particular simulation is vital if accurate results are to be obtained at a reasonable cost. Due to the complex structure of the vertebral body, it is difficult to mesh completely with hexahedrons. Therefore, tetrahedral elements may be necessary; the cancellous bone and the posterior part of the vertebral body are divided into 4-node tetrahedral elements, and the cortical bone and endplate are divided into 3-node shell elements. Components of each vertebra are show in Figure 1.

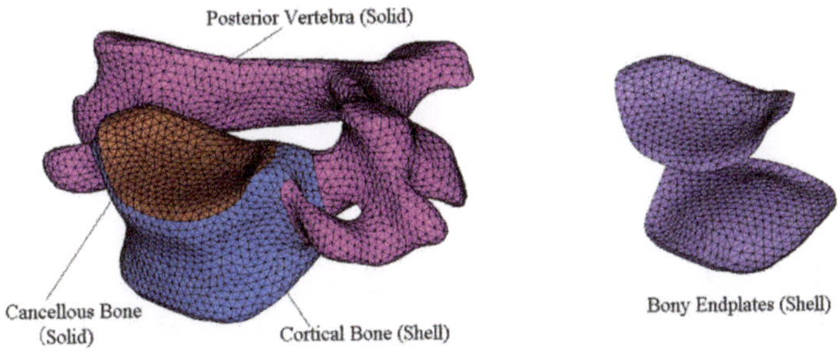

Figure 1. Components of each vertebra.

An intervertebral disc is a sealed body located between two adjacent vertebrae of the human spine and is composed of cartilaginous endplate, peripheral annulus fibrosus and central nucleus pulposus, in which the volume of nucleus pulposus accounts for about 44% of the volume of the whole intervertebral disc [16]. The annulus fibrosus adopts a four-layer hexahedral element from the outside to the inside, and a truss element with a cross-angle close to $\pm 35^\circ$ and only tension is used to simulate the annulus fibrosus fiber [29].

The ligaments were simulated using nonlinear connecting elements, which can simulate the nonlinear stretching process of ligaments based on the force-displacement data provided by the experimental data. This includes the anterior longitudinal ligament (ALL), posterior longitudinal ligament (PLL), ligamentum flavum, (FL), interspinous ligament (ISL) and capsular ligament (CL). The starting and ending positions of ligaments and the number of elements were determined according to the anatomical structure of cervical spine and related literature [30,31]. The contact behavior of the posterior facet joint was defined by using contact pairs with Abaqus 6.14 (Dassault Systèmes Simulia Corp., Providence, RI, USA). The finite-sliding, surface-to-surface formulation was used in this model, and the relationship between contact stress and clearance was set as an exponential. The whole finite element model included 75,648 nodes and 319,184 elements. The detailed data of element types and material properties are shown in Table 1, and the finite element model of segments C2-C7 is shown in Figure 2a. The detailed model of segments C5-C6 and the intervertebral disc model are shown in Figure 2b,c, respectively. After completing the finite element model using Hypermesh 13.0 software, the model was imported into Abaqus software. The solution and post-processing were completed with Abaqus software.

Table 1. Element types and mechanical properties of the FE model.

Component	Element Type	Young's Modulus (MPa)	Poisson's Ratio	Reference
Cortical bone	S3	12,000	0.29	[18]
Cancellous bone	C3D4	100	0.29	[32]
Posterior vertebra	C3D4	3500	0.29	[18]
Articular cartilage	C3D6	10.4	0.4	[33]
Endplate	S3	500	0.4	[18]
Nucleus pulposus	C3D8H	Hyperelastic (C10 = 0.348, D1 = 0.3)		[34]
Annulus ground substance	C3D8R	Hyperelastic (C10 = 0.12, C01 = 0.09)		[35]
Annulus fiber	T3D2	358–550	0.3	[36]
ALL	CONN3D2	Nonlinear (Incompressible)		[19,37]
PLL	CONN3D2	-		-
FL	CONN3D2	-		-
ISL	CONN3D2	-		-
CL	CONN3D2	-		-

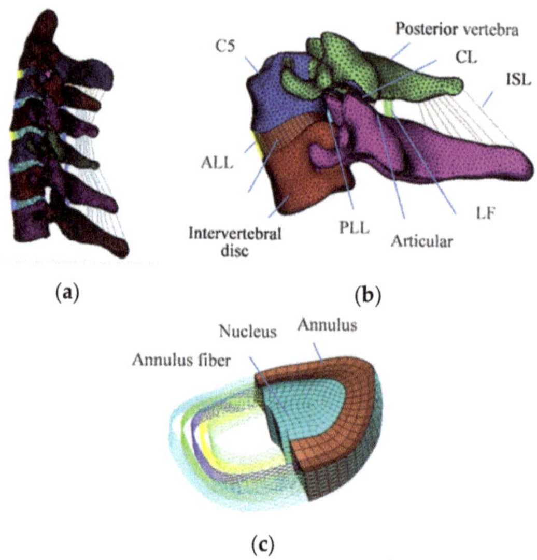

Figure 2. (**a**) Finite element model of segments C2-C7; (**b**) finite element model of segments C5-C6; (**c**) intervertebral disc model.

2.2. *Load and Boundary Conditions*

Binding constraints were set between the upper and lower surfaces of the intervertebral disc and the corresponding vertebral surfaces (endplates). Constraints were set on the 6 degrees of freedom of all nodes of the lower endplate of the C7 vertebral body. In order to compare with the results of previous in vitro experiments and verify the validity of the finite element model, the pure moment loads of 0.33 Nm, 0.5 Nm, 1 Nm, 1.5 Nm and 2 Nm were applied for flexion (+) and extension (−), and 1 Nm was applied for the left and right lateral bending and axial rotation. These loads also ensured that the range of motion of the cervical spine was within the normal range. The range of motion and intervertebral disc stress on each segment of the cervical spine under different load conditions were analyzed using finite element analysis.

3. Results

3.1. Range of Motion

By applying different flexion/extension loads to the model, the nonlinearity of cervical spine motion is well simulated. The range of motion during flexion is slightly larger than that during extension, which may be related to the effects of the facet joint on cervical spine motion during extension. In different segments, the results are shown in Figure 3. In flexion loading, the results of segments C2-C3, C4-C5 and C6-C7 were close to the values of Wheeldon et al.'s [3] and Camacho et al.'s [38] experimental data. In segments C3-C4 and C5-C6, the present study's results were close to the value of Camacho et al.'s [38] but stiffer than those of Wheeldon et al. [3] and Nightingale et al. [2], especially in segments C5-C6. In segments C6-C7, the results of the model were more flexible than those of Nightingale et al. [4] and Camacho et al. [38]. During the extension, the range of motion under different load conditions was close to the results of Wheeldon et al. [3], except in segments C2-C3, which was reduced by about 36% in comparison.

In Figures 4 and 5, the range of motion of lateral bending and axial rotation of different segments were compared with those of Panjabi et al. [1] and Zhang et al. [18]. In lateral bending, the results of segments C3-C4 and C4-C5 were smaller, different by about one standard deviation compared to those of Panjabi et al. [1]. In the rest of the segments (C2-C3, C5-C6 and C6-C7), the results and the mean value were in good agreement with the experimental data. In axial rotation, the mean value and the ranges of motion of segments C3-C4 and C5-C6 were close to the results of Panjabi et al. [1]. In segments C2-C3 and C6-C7, the finite element model was more flexible.

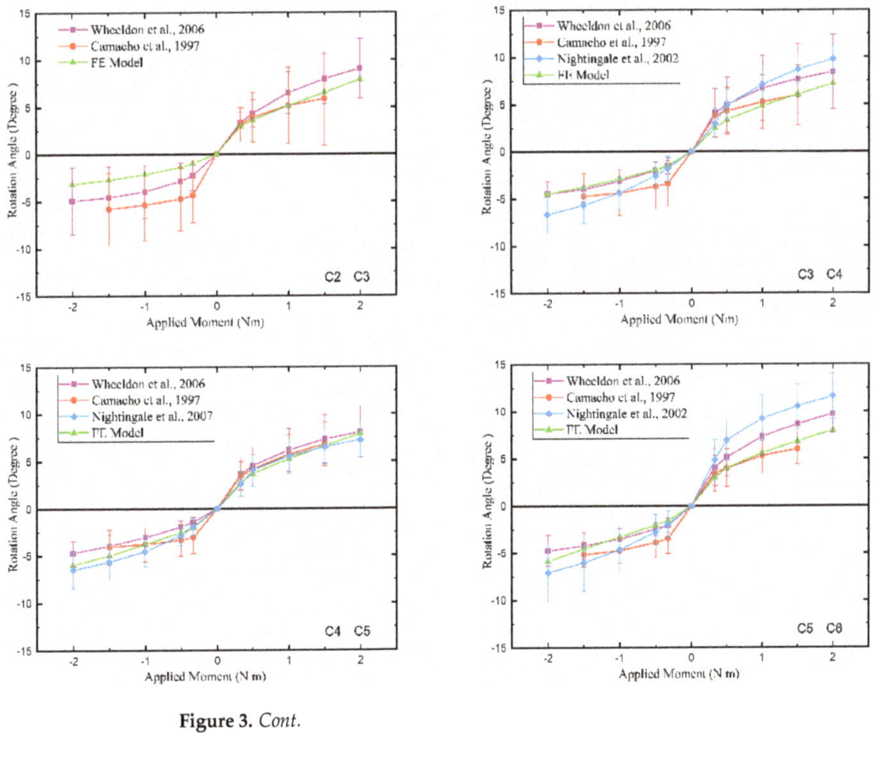

Figure 3. *Cont.*

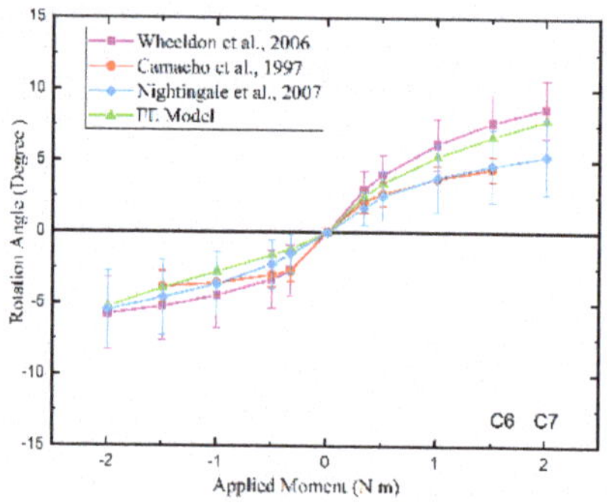

Figure 3. Comparison of the results of the FE model and experimental studies in different segments under flexion (+) and extension (−) (data from [3,4,38]).

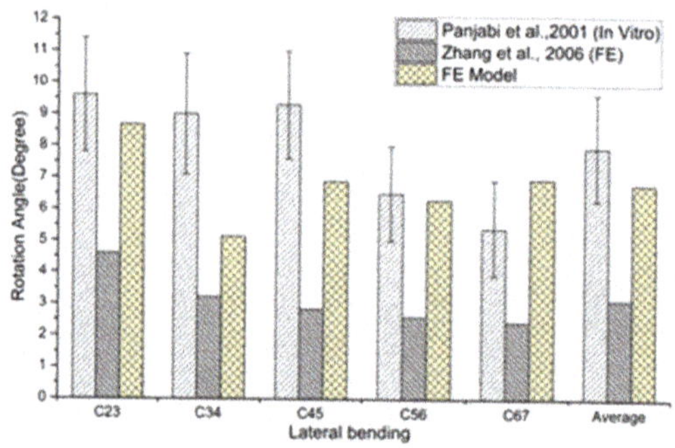

Figure 4. Comparison of ROM under lateral bending moment of 1.0 Nm (data from [1,18]).

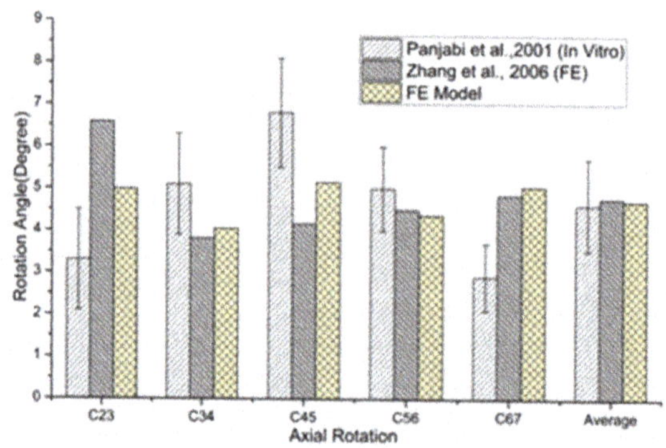

Figure 5. Comparison of ROM under axial rotation moment of 1.0 Nm (data from [1,18]).

3.2. *Intervertebral Disc Stress*

By applying different axial loads to simulate intervertebral disc compression, the intervertebral disc stress level and variation trend of segments C4-C5 are consistent with the relevant experimental results [39], as shown in Figure 6. In addition, in the process of simulating forward flexion/extension movement with a preloaded 10 N axial load, the stresses on the C3-C4 and C5-C6 intervertebral discs were smaller those the experimental results of Pospiech et al. [40]. As shown in Figure 7, the results of the finite element model were reduced by about 46% in segments C3-C4 and by about 35% in segments C5-C6 in comparison.

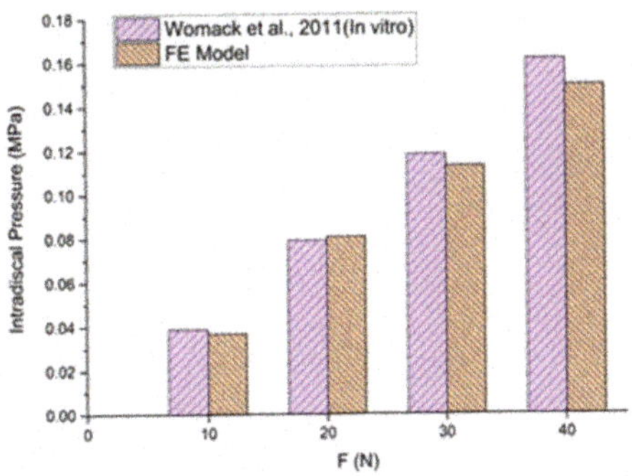

Figure 6. Intervertebral disc stress on segments C4-C5 under axial compression (data from [39]).

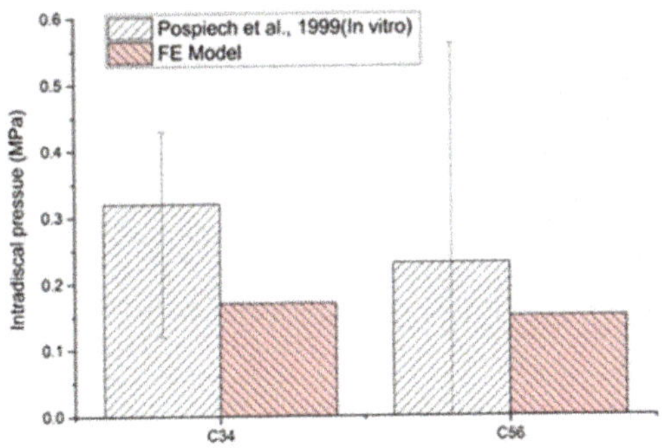

Figure 7. Intervertebral disc stress on segments C3-C4 and C5-C6 under flexion/extension moment of 0.5 Nm and pre-compression load of 10 N (data from [40]).

4. Discussion

By processing CT images to generate contour data of the vertebral body, the intervertebral disc was accomplished using general software to ensure the accuracy of the model and convenience for follow-up processing. This opens up a new way for the finite element analysis of complex structures. The selection of materials was further refined according to the current references and the actual conditions. For example, the past finite element model usually used uncompressed truss elements to simulate ligament tissues while the ligament shows a strong nonlinearity in the process of stretching [31,41]. Therefore, the nonlinear lconnecting element is used in this model to better present that feature. In addition, the construction of the intervertebral disc model also plays a key role in the whole finite element analysis. In this model, different hyperelastic materials were adopted to simulate the nucleus pulposus and the base of annulus fibrosus with five diagonally intersected layers of tensile-stressed truss elements from outside to inside with a $\pm 35°$ distribution [29], which guarantees the simulation of the intervertebral disc in both shape and structure.

Due to the restrictions on the in vitro experiments, our research group conducted an in-depth analysis of the relevant published in vitro experiments to verify the effectiveness of the model. A couple well-accepted classic experiments were selected for reference. Generally speaking, the range of motion and the variation trend resulting from the established finite element model of segments C2-C7 of the cervical vertebra are consistent with the results of the in vitro experiments in terms of flexion/extension, lateral bending and axial rotation. In the process of flexion, the finite element model analysis reproduces the nonlinearity of cervical spine movement while the results of different segments fall in the range of the different experimental results [2–4,38]. In the process of extension, the finite element model shows a sound coherence with the experimental results of Wheeldon [3] and Nightingale [2,4] in segments C3-C4, C4-C5 and C5-C6, with a generally higher stiffness. In segments C2-C3 and C6-C7, the stiffnesses of the finite element model is higher than those of the in vitro experiments. Although the results were within the range of a standard deviation of the experimental data, the range of motion of segments C2-C3 was reduced by 36% compared to the experimental mean of Wheeldon et al. [3] and by about 50% compared to the experimental mean of Camacho et al. [38]. We hold that the differences may be due to the thickness and size of the posterior articular cartilage and the direction of contact during extension. Relevant studies also showed that the orientation of the facet and the settings of the contact parameters could lead to different analysis results [36,37]. In terms of anatomical structure, the articular cartilage in C2-C3 and C6-C7 was thicker than those in the other segments [42], of which the difference was not fully reflected in the finite element modeling. In addition, to facilitate the finite element analysis (FEA) solution, a large contact range is preferred when setting the articular contact surface, which could be another cause for such results. In the process of lateral bending, the results of segments C2-C3 and C5-C6 meet the experimental results [1], whereas those of segments C3-C5 are smaller and those of segments C6-C7 are larger. Moreover, the ranges of motion of all segments are larger than the results of the finite element analysis from Zhang [18]. In the analysis of axial rotational movement, the results of this model are in correspondence with both the in vitro experiment from Panjabi [1] and the finite element analysis from Zhang [18]. It is also found in the analysis that the direction in which a moment of force is applied has a greater impact on the results of lateral bending and axial rotation than those of flexion and extension. Therefore, the load applied in the model verification is consistent with those of the in vitro experiments.

To further verify the effectiveness of the model, this research also targets intervertebral disc stress. However, as the cervical vertebra is smaller and more complicated than the lumbar vertebra in structure, this research only makes comparisons with the experimental results in terms of intervertebral disc stress on partial segments under two different loads. The results show that under the axial compressive load, the intervertebral disc stress level and the trend of segments C4-C5 correspond with the experimental results. The intervertebral disc stress on segments C3-C4 and C5-C6 under a flexion/extension moment of 0.5 Nm with a pre-compression load of 10 N is lower than the average of the experimental value. The intervertebral disc is an early and easily degenerated and aged organ in the human body. In Pospiech et al.'s [40] experiment, the mean age of the selected specimens was 44.3 (27 to 64); in the FEA, the material model could not fully simulate the actual situation, which might be the main cause for the differences.

5. Conclusions

The three-dimensional finite element model of segments C2-C7 is proved to be convenient and feasible through the comparison of the model with the in vitro experiments under different load conditions. The model analysis has shown coherence with the results of the in vitro experiments in terms of either the range of motion or intervertebral disc stress. The model can be applied to analyze the biomechanical characteristics of the cervical spine, providing a possibility for the study of the mechanism for cervical spine disease as well as prosthesis implantation performance. However, due to the complex structure of cervical spine, there are large deformations, material nonlinearity and contact nonlinearity in the process of analysis. Although this model tries to make the setting of the shape and material properties close to reality, some simplifications are inevitably adopted. The influence of muscles, blood vessels and the spinal cord was ignored in the process of establishing the finite element model, which still needs to be further improved in future analysis.

Author Contributions: X.C., T.W. and C.P. contributed substantially to the conception and design of the experiments. X.C. conducted the experiments and wrote the manuscript. T.W. and C.P. conducted the data analyses. All authors have read and agreed to the published version of the manuscript.

Funding: This research was funded by the International Science and Technology Cooperation Programme (2018YFE0194100), the National Natural Science Foundation of China (31870952) and the Priority Academic Program Development of Jiangsu Higher Education Institutions.

Institutional Review Board Statement: Not applicable.

Informed Consent Statement: Not applicable.

Data Availability Statement: The original contributions presented in this study are included in the article; further inquiries can be directed to the corresponding author.

Conflicts of Interest: The authors declare no conflict of interest.

References

1. Panjabi, M.; Crisco, J.J.; Vasavada, A.; Oda, T.; Cholewicki, J.; Nibu, K.; Shin, E. Mechanical properties of the human cervical spine as shown by three-dimensional load-displacement curves. *Spine* **2001**, *26*, 2692–2700. [CrossRef]
2. Nightingale, R.W.; Winkelstein, B.A.; Knaub, K.E.; Richardson, W.J.; Luck, J.F.; Myers, B.S. Comparative strengths and structural properties of the upper and lower cervical spine in flexion and extension. *J. Biomech.* **2002**, *35*, 725–732. [CrossRef]
3. Wheeldon, J.A.; Pintar, F.A.; Knowles, S.; Yoganandan, N. Experimental flexion/extension data corridors for validation of finite element models of the young, normal cervical spine. *J. Biomech.* **2006**, *39*, 375–380. [CrossRef]
4. Nightingale, R.W.; Carol Chancey, V.; Ottaviano, D.; Luck, J.F.; Tran, L.; Prange, M.; Myers, B.S. Flexion and extension structural properties and strengths for male cervical spine segments. *J. Biomech.* **2007**, *40*, 535–542. [CrossRef]
5. Liu, Q.; Guo, Q.; Yang, J.; Zhang, P.; Xu, T.; Cheng, X.; Chen, J.; Guan, H.; Ni, B. Subaxial Cervical Intradiscal Pressure and Segmental Kinematics Following Atlantoaxial Fixation in Different Angles. *World Neurosurg.* **2016**, *87*, 521–528. [CrossRef]
6. Belytschko, T.; Kulak, R.F.; Schultz, A.B.; Galante, J.O. Finite element stress analysis of an intervertebral disc. *J. Biomech.* **1974**, *7*, 277. [CrossRef]
7. Bozic, K.J.; Keyak, J.H.; Skinner, H.B.; Bueff, H.U.; Bradford, D.S. Three-dimensional finite element modeling of a cervical vertebra: An investigation of burst fracture mechanism. *J. Spinal Disord.* **1994**, *7*, 102–110. [CrossRef]
8. Teo, E.C.; Paul, J.P.; Evans, J.H. Finite element stress analysis of a cadaver second cervical vertebra. *Med. Biol. Eng. Comput.* **1994**, *32*, 236.
9. Teo, E.C.; Ng, H.W. Evaluation of the Role of Ligaments, Facets and Disc Nucleus in Lower Cervical Spine Under Compression and Sagittal Moments Using Finite Element Method. *Med. Eng. Phys.* **2001**, *23*, 155–164. [CrossRef]
10. Del Palomar, A.P.; Calvo, B.; Doblaré, M. An Accurate Finite Element Model of the Cervical Spine Under Quasi-Static Loading. *J. Biomech.* **2008**, *41*, 523–531. [CrossRef]
11. Panzer, M.B.; Fice, J.B.; Cronin, D.S. Cervical Spine Response in Frontal Crash. *Med. Eng. Phys.* **2011**, *33*, 1147–1159. [CrossRef] [PubMed]
12. Oxland, T.R. Fundamental Biomechanics of the Spine—What we Have Learned in the Past 25 Years and Future Directions. *J. Biomech.* **2016**, *49*, 817–832. [CrossRef]
13. Kim, Y.H.; Khuyagbaatar, B.; Kim, K. Recent Advances in Finite Element Modeling of the Human Cervical Spine. *J. Mech. Sci. Technol.* **2018**, *32*, 1–10. [CrossRef]
14. DeWit, J.A.; Cronin, D.S. Cervical Spine Segment Finite Element Model for Traumatic Injury Prediction. *J. Mech. Behav. Biomed.* **2012**, *10*, 138–150. [CrossRef] [PubMed]
15. Cronin, D.S. Finite Element Modeling of Potential Cervical Spine Pain Sources in Neutral Position Low Speed Rear Impact. *J. Mech. Behav. Biomed.* **2014**, *33*, 55–66. [CrossRef] [PubMed]
16. Mustafy, T.; Moglo, K.; Adeeb, S.; El-Rich, M. Injury Mechanisms of the Ligamentous Cervical C2-C3 Functional Spinal Unit to Complex Loading Modes: Finite Element Study. *J. Mech. Behav. Biomed.* **2016**, *53*, 384–396. [CrossRef] [PubMed]
17. Meyer, F.; Humm, J.; Purushothaman, Y.; Willinger, R.; Pintar, F.A.; Yoganandan, N. Forces and Moments in Cervical Spinal Column Segments in Frontal Impacts Using Finite Element Modeling and Human Cadaver Tests. *J. Mech. Behav. Biomed.* **2019**, *90*, 681–688. [CrossRef]
18. Zhang, Q.H.; Teo, E.C.; Ng, H.W.; Lee, V.S. Finite Element Analysis of Moment-Rotation Relationships for Human Cervical Spine. *J. Biomech.* **2006**, *39*, 189–193. [CrossRef]
19. Kallemeyn, N.; Gandhi, A.; Kode, S.; Shivanna, K.; Smucker, J.; Grosland, N. Validation of a C2–C7 Cervical Spine Finite Element Model Using Specimen-Specific Flexibility Data. *Med. Eng. Phys.* **2010**, *32*, 482–489. [CrossRef]
20. Erbulut, D.U.; Zafarparandeh, I.; Lazoglu, I.; Ozer, A.F. Application of an Asymmetric Finite Element Model of the C2-T1 Cervical Spine for Evaluating the Role of Soft Tissues in Stability. *Med. Eng. Phys.* **2014**, *36*, 915–921. [CrossRef] [PubMed]
21. Herron, M.R.; Park, J.; Dailey, A.T.; Brockmeyer, D.L.; Ellis, B.J. Febio Finite Element Models of the Human Cervical Spine. *J. Biomech.* **2020**, *113*, 110077. [CrossRef] [PubMed]

22. Hua, W.; Zhi, J.; Ke, W.; Wang, B.; Yang, S.; Li, L.; Yang, C. Adjacent Segment Biomechanical Changes After One- Or Two-Level Anterior Cervical Discectomy and Fusion Using Either a Zero-Profile Device or Cage Plus Plate: A Finite Element Analysis. *Comput. Biol. Med.* **2020**, *120*, 103760. [CrossRef] [PubMed]
23. Chen, C.; Yuchi, C.X.; Gao, Z.; Ma, X.; Zhao, D.; Li, J.W.; Xu, B.; Zhang, C.Q.; Wang, Z.; Du, C.F.; et al. Comparative Analysis of the Biomechanics of the Adjacent Segments After Minimally Invasive Cervical Surgeries Versus Anterior Cervical Discectomy and Fusion: A Finite Element Study. *J. Orthop. Transl.* **2020**, *23*, 107–112. [CrossRef] [PubMed]
24. Khalaf, K.; Nikkhoo, M. Comparative Biomechanical Analyses of Lower Cervical Spine Post Anterior Fusion Versus Intervertebral Disc Arthroplasty: A Geometrically Patient-Specific Poroelastic Finite Element Investigation. *J. Orthop. Transl.* **2022**, *36*, 33–43. [CrossRef]
25. Nishida, N.; Tripathi, S.; Mumtaz, M.; Kelkar, A.; Kumaran, Y.; Sakai, T.; Goel, V.K. Soft Tissue Injury in Cervical Spine is a Risk Factor for Intersegmental Instability: A Finite Element Analysis. *World Neurosurg.* **2022**, *164*, e358–e366. [CrossRef]
26. Yuan, W.; Zhang, H.; Zhou, X.; Wu, W.; Zhu, Y. The Influence of Artificial Cervical Disc Prosthesis Height on the Cervical Biomechanics: A Finite Element Study. *World Neurosurg.* **2018**, *113*, e490–e498. [CrossRef]
27. Zhou, E.; Huang, H.; Zhao, Y.; Wang, L.; Fan, Y. The Effects of Titanium Mesh Cage Size on the Biomechanical Responses of Cervical Spine After Anterior Cervical Corpectomy and Fusion: A Finite Element Study. *Clin. Biomech.* **2022**, *91*, 105547. [CrossRef]
28. Wang, K.; Wang, H.; Deng, Z.; Li, Z.; Zhan, H.; Niu, W. Cervical Traction Therapy with and without Neck Support: A Finite Element Analysis. *Musculoskel. Sci. Prac.* **2017**, *28*, 1–9. [CrossRef]
29. Schmidt, H.; Heuer, F.; Drumm, J.; Klezl, Z.; Claes, L.; Wilke, H.J. Application of a calibration method provides more realistic results for a finite element model of a lumbar spinal segment. *Clin. Biomech.* **2007**, *22*, 377–384. [CrossRef]
30. Panjabi, M.M. Cervical Spine Models for Biomechanical Research. *Spine* **1998**, *23*, 2684. [CrossRef]
31. Yoganandan, N.; Kumaresan, S.; Pintar, F.A. Geometric and mechanical properties of human cervical spine ligaments. *J. Biomech. Eng.* **2000**, *122*, 623. [CrossRef] [PubMed]
32. Ha, S.K. Finite element modeling of multi-level cervical spinal segments (c3–c6) and biomechanical analysis of an elastomer-type prosthetic disc. *Med. Eng. Phys.* **2006**, *28*, 534–541. [CrossRef]
33. Kumaresan, S.; Yoganandan, N.; Pintar, F.A. Finite element modeling approaches of human cervical spine facet joint capsule. *J. Biomech.* **1998**, *31*, 371–376. [CrossRef] [PubMed]
34. Li, Y.; Lewis, G. Influence of the constitutive material behavior model assigned to the annulus fibrosus and the nucleus pulposus on the biomechanical performance of a model of the cervical spine: A finite element analysis study. *J. Mech. Med. Biol.* **2010**, *10*, 151–166. [CrossRef]
35. Ezquerro, F.; Vacas, F.G.; Postigo, S.; Prado, M.; Simón, A. Calibration of the finite element model of a lumbar functional spinal unit using an optimization technique based on differential evolution. *Med. Eng. Phys.* **2011**, *33*, 89–95. [CrossRef]
36. Kim, H.; Chun, H.; Lee, H.; Kang, K.; Lee, C.; Chang, B.; Yeom, J.S. The biomechanical influence of the facet joint orientation and the facet tropism in the lumbar spine. *Spine J.* **2013**, *13*, 1301–1308. [CrossRef] [PubMed]
37. Panzer, M.B.; Cronin, D.S. C4–c5 segment finite element model development, validation, and load-sharing investigation. *J. Biomech.* **2009**, *42*, 480–490. [CrossRef] [PubMed]
38. Camacho, D.L.; Nightingale, R.W.; Robinette, J.J.; Vanguri, S.K.; Coates, D.J.; Myers, B.S. (Eds.) *Experimental Flexibility Measurements for the Development of a Computational Head-Neck Model Validated for Near-Vertex Head Impact*; SAE International: Warrendale, PA, USA, 1997.
39. Womack, W.; Leahy, P.D.; Patel, V.V.; Puttlitz, C.M. Finite element modeling of kinematic and load transmission alterations due to cervical intervertebral disc replacement. *Spine* **2011**, *36*, E1126–E1133. [CrossRef] [PubMed]
40. Pospiech, J.; Stolke, D.; Wilke, H.J.; Claes, L.E. Intradiscal pressure recordings in the cervical spine. *Neurosurgery* **1999**, *44*, 379–385. [CrossRef]
41. Mattucci, S.F.E.; Moulton, J.A.; Chandrashekar, N.; Cronin, D.S. Strain rate dependent properties of younger human cervical spine ligaments. *J. Mech. Behav. Biomed.* **2012**, *10*, 216–226. [CrossRef]
42. Yoganandan, N.; Knowles, S.A.; Maiman, D.J.; Pintar, F.A. Anatomic Study of the Morphology of Human Cervical Facet Joint. *Spine* **2003**, *28*, 2317–2323. [CrossRef] [PubMed]

metals

Review

Advances in Schiff Base and Its Coating on Metal Biomaterials—A Review

Zhiqiang Zhang [1,†], Qingya Song [1,†], Yubin Jin [1], Yashan Feng [2], Jingan Li [1,*] and Kun Zhang [3,*]

1 School of Materials Science and Engineering, Zhengzhou University, Zhengzhou 450001, China
2 Zhengzhou Railway Vocational & Technical College, Zhengzhou 450000, China
3 School of Life Science, Zhengzhou University, Zhengzhou 450001, China
* Correspondence: lijingan@home.swjtu.edu.cn (J.L.); zhangkun@zzu.edu.cn (K.Z.); Tel.: +86-185-3995-6211 (J.L.)
† Joint first authors.

Abstract: In recent years, metal biomaterials have emerged one after another, and have many excellent properties, playing a great role in medicine. However, these coatings cannot meet the medical needs in every aspect. Schiff base is an important organic synthetic reagent and liquid crystal material in organic chemistry. It mainly refers to a class of organic compounds containing imine or azomethine characteristic groups (-RC=N-). It has important anti-tumor, anti-virus, antifungal and antibacterial activities. Based on the excellent properties of Schiff base, the coatings made of Schiff base can improve the bioactivity of materials, which have a good development prospect in medicine. In this paper, the preparation methods and properties of Schiff base and many advantages of Schiff base coatings are reviewed. The research on the modification of coatings or functional membranes by Schiff base and Schiff base reaction, as well as the extensive application of special Schiff base coatings in many fields such as anti-corrosion, antibacterial, flame retardant, etc., are carried out. Suggestions for further research on Schiff base coatings on metal biomaterials are put forward.

Keywords: metal biomaterials; Schiff base coatings; corrosion resistance; biocompatibility

check for updates

Citation: Zhang, Z.; Song, Q.; Jin, Y.; Feng, Y.; Li, J.; Zhang, K. Advances in Schiff Base and Its Coating on Metal Biomaterials—A Review. *Metals* **2023**, *13*, 386. https://doi.org/10.3390/met13020386

Academic Editor: Mosab Kaseem

Received: 22 December 2022
Revised: 6 February 2023
Accepted: 10 February 2023
Published: 13 February 2023

1. Introduction

Biomaterials are materials used to repair or replace diseased or damaged parts of the human body. The importance of biomaterials has increased significantly with population growth and aging, as it contributes to improving the quality of life and prolonging life spans. The surface structure of biomaterials is the key to control their interaction with the biological environment [1]. In recent years, metal biomaterials have been widely used in medical treatment, and various coatings have also brought great improvement to the functions of materials. So far, high requirements have been proposed for the preparation of uniform and corrosion resistant metallic material coatings [2].

Using the composite technology of different kinds of conventional materials to generate new practical functions is one of the strategies to accelerate the development of the next generation of biomaterials [3]. Schiff bases are a class of important bases in organic chemistry, with anti-tumor, anti-virus, antifungal and antibacterial activities [4]. Biomaterials can be ceramics, polymers or metals. Although ceramics and polymers are widely used in various medical applications, metal implants still occupy a dominant position because of their advantages, such as ease of manufacture [5]. In recent years, it has been applied in medicine and other fields, and has good prospects for development. This paper introduces Schiff base and its advantages as coatings.

In some environments, many biological macromolecules will be decomposed, but the synthetic biological composites are stable [6]. The biocompatibility of materials depends on their surface characteristics, such as chemical composition, charge density, wettability and topography, which directly affect tissue reactions, such as inflammatory reaction and tissue regeneration [7]. Among metal materials, 316L stainless steel is widely used as a temporary biological material due to its excellent mechanical properties, sufficient biocompatibility, high corrosion resistance, low inherent toxicity, availability and low economic cost. However, the interface between metal materials and electrolytes is prone to electrochemical corrosion [8], which has a great impact on the performance and service life of metals. In order to solve these problems, scientists have studied various types of coatings for metal surfaces. A modified coating composed of two Schiff base complexes ($Cr^{3+}$and Ni^{3+}) and SiO_2 nanoparticles was constructed, which greatly enhanced the protective property and mechanical resistance of carbon steel on the surface of carbon steel [9].

The coatings on these metal surfaces not only beautify materials, but also reduce corrosion and wear of materials. Therefore, modern research mainly focuses on anti-corrosion coatings on metal surfaces [10]. However, corrosive substances cannot be completely blocked by the coating, and more or less corrosive substances will enter the metal/coating interface. Therefore, people try to mix and improve the coating in a variety of ways to provide better shielding against corrosive media [8]. In general, corrosion inhibition mainly depends on the physical and chemical properties of the inhibitor molecule, the π orbital characteristics of the electron donor and the electronic structure of the inhibitor. Schiff bases have been reported as useful corrosion inhibitors in metal and metal alloy corrosion media, which can be easily synthesized from relatively cheap materials. However, it is well known that the behavior of inhibitors depends on the interaction between functional groups in their molecules and metal surfaces. The inhibition efficiency of Schiff base is attributed to the existence of (-RC=N-) group and the electronegativity of any other heteroatoms (such as nitrogen, sulfur and oxygen) in the molecule. The high corrosion inhibition is caused by the lone pair electrons of N and the electronic distribution of the double bond of the imine unit [11]. Schiff base can stabilize the corrosion inhibition of steel through its good solubility and polar nitrogen atom, and with the increase of Schiff base ligand concentration, the protection ability of steel in corrosive media is improved [12]. The study of Schiff base coating is helpful to further explore the potential role and value of Schiff base. In this study, we will review the advances in Schiff base and its coating for the metal biomaterials application (Figure 1).

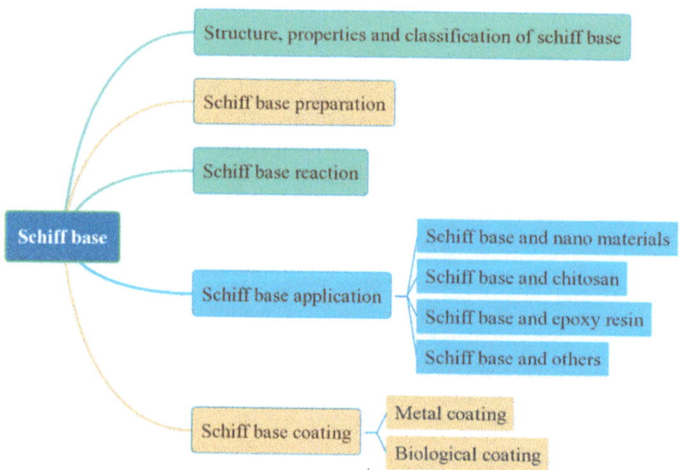

Figure 1. Summary of Schiff bases in this review.

2. Structure, Properties and Classification of Schiff Bases

Schiff bases are called "privileged ligands". They can bind with almost all metal ions and stabilize the oxidation state of metal ions under various useful conditions. This is why most metal ion Schiff base complexes have high activity. Schiff bases are also an important class of bases in organic chemistry. They have been widely studied because of their synthetic flexibility, selectivity, sensitivity to central metal atoms, structural similarity with natural biological compounds, and the presence of imino (-N-CH-) groups [4]. Schiff base with a methylene imino structure is widely used in many biological activities and industrial applications, such as antibacterial, anti-tumor, antioxidant, anti-inflammatory, antifungal, etc. Schiff base ligands are widely used in the development of inorganic chemistry and coordination chemistry because they can form complexes with metal ions. Moreover, Schiff bases will not change their original performance due to high temperature and pressure. Some Schiff bases show good catalytic performance at high temperature, which makes the application fields and researchability of Schiff bases more extensive.

Schiff base ligands, as bidentate compounds, coordinate with metal ions through methylene nitrogen and phenoxy centers in the form of single deprotonation [13]. Derivatives of Schiff bases and their complexes have also been proved to have many functions. Schiff base transition metal complexes are considered to be a kind of coordination compound with great potential. In recent years, they have been widely used in biochemistry, analysis, antibacterial agents, and other fields, especially in pharmaceutical chemistry, because of their advantages in antibacterial effect [14]. In addition, because the Schiff base of salicylaldehyde and its derivatives can coordinate with some metal ions, especially copper ions, salicylaldehyde assembled on the electrode surface by Schiff base reaction can be used as a copper ion sensor and its related applications, such as electrocatalysis of benzoquinone [15].

Schiff bases are easy-to-synthesize ligands formed by condensation of aldehydes and imines. They are often used as efficient catalysts for various reactions in homogeneous and heterogeneous catalytic reactions. For example, the Schiff-base-mediated gold nano catalyst directly catalyzes the hydrogenation of carbon dioxide to generate formate [16]. In recent years, the self-healing properties of Schiff bases have received extensive attention, but many self-healing reactions require external stimuli, such as light, heat, and pH. The Schiff base bond is one of the few covalent bonds which can repair itself without external stimulation. In addition, when exposed to a liquid environment, water can promote the kinetic properties of Schiff base reactions [17]. Schiff bases are a class of important bases in organic chemistry, with anti-tumor, anti-virus, antifungal, and antibacterial activities [4]. Schiff bases are a class of very famous organic molecules with a variety of structures and properties, as displayed in Table 1 [18].

Table 1. Classification of Schiff base [18].

Type	Examples
Bidentate	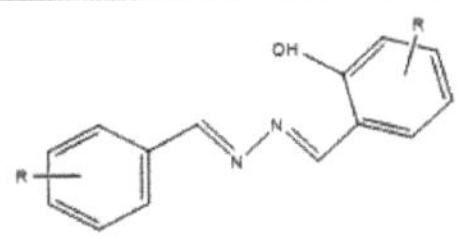
Tridentate	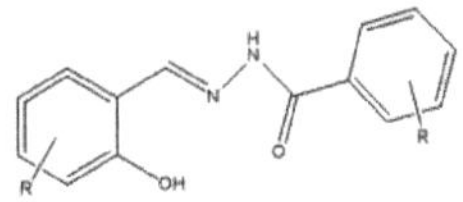
Tetradentate	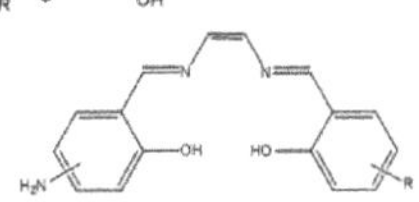
Polydentate	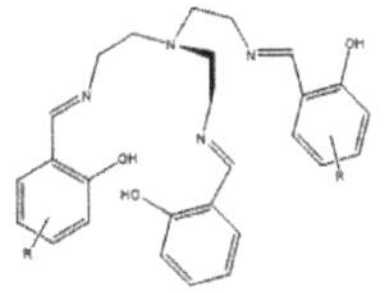

Another key point to note is that Schiff bases containing oxygen, nitrogen and carbonyl groups have been used as drugs. It is reported that Schiff bases have biological activity against bacteria and fungi due to their biochemical, clinical and pharmacological characteristics. The bioactivity of Schiff base mainly depends on the imine group. Therefore, nitrogen atoms may participate in the process of forming hydrogen bonds with active centers of cell components and interfere with normal cell processes [19]. In addition, the unique molecular structure of Schiff base compounds can be used to modify coatings on metals to obtain better physical and chemical properties [20].

3. Synthesis of Schiff Base

The synthesis of Schiff bases is catalyzed by various types of chemical catalysts [21]. The imino group is considered as the basic characteristic of Schiff base, which has interesting biological significance and is found to be responsible for biological activities, such as bactericidal activities. Because the structural characteristics of Schiff bases provide target-specific recognition sites, the combination of Schiff bases and nanomaterials shows synergistic benefits, which has attracted extensive attention from researchers [18].

Gossypol is an effective enantiomer and has good therapeutic prospects for various types of cancer. Racemic gossypol can be separated from cottonseed, and its enantiomer can be purified by using the diastereomeric resolution technology of methyl tryptophan hydrochloride. This method is environmentally friendly and will not produce many pollutants. Schiff base can also be obtained by modifying high molecular weight and low molecular weight chitosan with phenol condensation reaction [22].

The synthesis of Schiff bases requires the presence of organic bases as catalysts. FeCu@N-doped carbon is an efficient catalyst, which can be used to convert amines and alcohols into Schiff bases. Bifunctional cobalt/zinc mesoporous silica nanoparticles can also be used to synthesize Schiff bases from aromatic amines and benzyl alcohols. The reaction temperature was 120 °C, and the reaction was carried out in the presence of air flow for 3 h. The mixture of P_2O_5 and Al_2O_3 was used to catalyze the synthesis of Schiff base from carbonyl compounds and primary amines under solvent-free conditions [4].

Many of the most abundant polysaccharides based on monosaccharides can be simply modified with a large number of aldehyde groups, such as dialdehyde cellulose (DAC). Due to the existence of a large number of active aldehyde groups on the skeleton, these modified polysaccharides can be used to prepare Schiff base complexes [23]. In addition, polyshev bases can be prepared by a solvent-free method. A mixture of bifurfural (0.526 mmol) and diamine (0.526 mmol) was stirred at 140 °C in a 25 mL double neck flask under nitrogen for 3 h. After the reaction mixture was heated under vacuum at 140 °C for 3 h, the solid obtained by washing with methanol was heated under vacuum at 80 °C to remove residual methanol. The resulting solids were collected as yellowish brown solids or reddish brown materials [24]. Common preparation method of Schiff base: in methanol medium, the mixed solution is stirred continuously for 2 h with a magnetic stirrer at 80 °C, cooled to separate yellow orange crystal products, washed with cold methanol and dried, and the purity is identified by thin layer analysis [25].

Toluidine (0.01 mol) was mixed with azo coupled 4—(3-formyl-4-hydroxy radical) diazoyl-n-5 (5-methylisoxazol-3-yl) benzenesulfonamide (0.01 mol) in equal molar ratio (1:1). The reaction mixture was mixed with 3~4 mL methanol and irradiated in a microwave oven. The reflection time was 4 min and the microwave power was 90 w. Orange products were recrystallized with hot ethanol, and finally dried with anhydrous $CaCl_2$ in a dryer under reduced pressure (pollution-free, environmental protection, low cost, high yield, simple operation) [26]. In addition, in some experiments, the Schiff base was synthesized with 0.2 mol/L L-cysteine and 0.2 mol/L glucose [20].

4. Schiff Base Reaction

The Schiff base reaction is widely used in chemistry because of its mild reaction conditions and high reaction rate. In recent years, Schiff base complexes have attracted extensive attention in biochemistry and biomedicine due to their unique properties [15]. The Schiff base reaction is a feasible surface modification strategy, because it has many advantages: (1) mild reaction conditions; (2) simple process; (3) the by-products are harmless; (4) high yield; and (5) the product is easy to separate [18].

The Schiff base reaction was proposed by German chemist Hugo Schiff in 1864. It refers to the reaction between a class of compounds containing aldehydes (or ketones) and amino groups to produce imino groups (-C=N-), which enable them to simultaneously have fluorescence emission and bonding ability [27]. Schiff bases and their derivatives can be easily synthesized through simple chemical reactions, which have better inhibition on metals and alloys than other organic molecules in acidic media [28].

First of all, Schiff base bonding makes the films have high stability and able to withstand harsh conditions. Secondly, the formation of in situ Schiff base avoids the additional post-treatment process and improves the stability of the multilayer. Third, the Schiff base reaction can be carried out in both aqueous and organic solutions. This allows the materials to be incorporated, dissolved only or used only in nonaqueous solutions. Fourth, the Schiff base reaction leaves redundant active groups (aldehydes and amino groups) in the multilayer, which can further react with other substances to adjust the product characteristics to adapt to different applications [15].

5. The Application of Schiff Base on Biomaterials

5.1. Schiff Base and Nano Materials

In recent years, the cross application of Schiff base chemistry and nanomaterials has become increasingly widespread. Nanomaterials have unique physical and chemical properties, and can play a synergistic role with Schiff base to prepare modified electrodes. The reasonable design of Schiff base structure can improve the selectivity to specific ions, and can be used as an excellent ionic group to enhance the electrochemical response of the electrode. In addition, Schiff base nano sensors are widely used in biology, industry, environment, food and other fields [18]. Polymer Schiff base with adjustable redox voltage is considered as a promising negative electrode material for sodium ion batteries [29].

The Schiff base reaction is used to prepare large-area polymer nanofilm (PTF) with adjustable thickness. The PTF contains a polymer chain network of conjugated Schiff bases, provides micropores, and is decorated with a combination bag arranged in space, which may adsorb uranium in octahedral geometry [30].

Schiff bases can also stabilize the activity of several metals in various oxidation states, so as to control the activity change of metals in the catalytic conversion process. It is also widely used in industry, including as a dye and pigment [31]. The new modified Schiff base nanoparticles obtained by crosslinking chitosan with bis (4-formyl-2-methoxyphenyl carbonate) have free radical scavenging ability and anti-cancer properties. The carbonate skeleton and part of the nanoparticles are hydrolyzed under the targeted carcinogenic microenvironment, releasing vanillin and chitosan, and enhancing the anti-cancer activity [32]. Chemical modification of chitosan with aromatic aldehydes such as vanillin, salicylaldehyde, and provanillin can prepare stable Schiff base ligands [33].

Polylysine and catechol can also be modified by the Schiff base reaction in one step, improving the adhesion of polylysine to inorganic particles and surfaces [34]. Dopa catechol can be prepared by Schiff base reaction. Its binding with metal ions is site specific. This preparation method is inspired by the formation of the cuticle before and after the formation of mussels, and a hard but scalable metal surface cross-linked protein composite coating with layered structure is prepared [35].

5.2. Schiff Base and Chitosan

Chitosan is a common natural polysaccharide, which is widely distributed in the skeleton of insects and crustaceans in the form of acetyl (chitin). The hydroxyl and amino groups (nucleophilic groups) on the chitosan chain provide suitable positions for various attractive chemical transformations. Although chitosan exists widely and its cost is low, it has moderate thermodynamic properties in epoxy ammonia system. The preparation of biological aromatic amines by Schiff base reaction and structure is a good solution under the premise of simple synthesis and renewability. Among them, polyelectrolyte chitosan Schiff bases (PECSBs) are very effective as antifouling agents at high levels of eutrophication, and have strong inhibitory effects on the sedimentation of brown algae, green algae and red algae [36].

Chitosan has different functional groups, has a good structure for preparing new derivatives [37], and a variety of magnetic adsorbents [38]. Carboxymethylation of chitosan can not only improve the hydrophobicity of polymers, but also eliminate the problem of water solubility [39]. Glycans and their derivatives have some unique properties in the field of pharmacy and pharmaceutical chemistry [40]. Chitosan derivatives are widely used in key pharmaceutical compounds due to their non-toxic and biodegradable properties. Chitosan Schiff base and its metal complexes can improve the biological activity of chitosan [41]. Chitosan Schiff base containing aromatic aldehyde is of great importance to the biology of pathogenic bacteria, fungi and multi-drug-resistant bacteria. It is reported that the biological potential of chemically modified chitosan in antibacterial and antifungal effects has been improved. Chitosan Schiff base fused with medicinal plants has a good inhibitory effect on experimental pathogens and multi-drug-resistant bacteria [42]. In addition, graphene oxide can be linked to chitosan through ester bonds to further modify chitosan Schiff base. This modification increases the number of metal ion binding sites [43].

The accurate determination of chitosan content is of great significance to the quality control of chitosan, so people pay more and more attention to it. Two chitosan Schiff base derivatives, chitosan-Schiff base bearing benzaldehyde (BCSB) and chitosan-Schiff base derivatives bearing propionaldehyde (PCSB) were synthesized according to the average degree of deacetylation of chitosan with benzaldehyde and propionaldehyde. The total mass of glucosamine hydrochloride obtained by hydrolysis of all Schiff base products was calculated, and then the theoretical mass of chitosan was deduced through reverse calculation. Finally, the sample mass of the Schiff base reaction was combined with the theoretical mass of chitosan to obtain the content of chitosan. This method is accurate and simple, and it provides a good idea and method for the determination of chitosan content [44].

An environment-friendly Schiff base, salicylaldehyde chitosan Schiff base, was synthesized from chitosan and salicylaldehyde. It can be adsorbed on the steel surface in the form of protonation as an inhibitor, and it is a new corrosion inhibitor in the oil and gas industry [45]. As a new adsorbent, Schiff base grafted graphene oxide magnetic chitosan was used to extract and quantify lead ions in blood samples by dispersive magnetic solid phase extraction [46]. Chitosan-coated manganese Schiff base complex supported on magnetic iron oxide nanoparticles is a promising nano catalyst [47].

Vanillin and trans cinnamaldehyde combine with chitosan through the Schiff base reaction and reductive amination reaction. It was used as an active coating on the surface of fresh cut muskmelon. The modified polysaccharide produces a stable and good adhesive coating, which does not affect the appearance of the product, and can be used as an active edible coating and a transport carrier directly used in food [48].

Cellulose-based materials are widely used in the biomedical field. However, the lack of antibacterial activity of cellulose fibers will inevitably lead to the damage of journals in preventing bacterial infection. In this study, cellulose Schiff base ligands were prepared by selective oxidation of cellulose fibers, introduction of aldehyde groups, and the Schiff base reaction of aldehyde groups and amino groups on glycine. A new type of cellulose Schiff base copper complex was synthesized by complexing copper ions with Schiff base ligands. The prepared composite greatly enhances the antibacterial property of cellulose fiber [49].

According to the research, researchers have prepared a new injectable hydrogel with biocompatibility and biodegradability through the Schiff base reaction, which is used for soft tissue adhesion, hemostasis and bone repair materials [50]. This study provides a strategy for the design and preparation of rapid in situ forming hydrogels. Through the in situ formation reaction of Schiff base, hydrogels extracted from natural polysaccharides can be modulated and prepared for biomedical applications, such as soft tissue adhesives and hemostatic agents in the future [51].

Double Schiff base bonds can restrict the intramolecular movement of single cluster horizontal surface motifs. Double Schiff base linkage can induce high brightness luminescence of gold nanoclusters at single cluster level in aqueous solution. Dialdehydes are commonly used to construct different building units of nanoscale structures with high precision and good controllability. Dialdehydes react with amino groups to form imines, which induce the crosslinking of each component. In conclusion, dialdehyde mediated cross-linking strategy is particularly attractive for water-soluble gold nano carbon [52].

Schiff base compounds used in the field of flame retardancy have also received extensive attention because they can cross link and generate nitrogen containing hexatomic rings, which is conducive to the formation of a stable cross-linking network at high temperatures, thus producing high residual carbon and giving it inherent flame retardancy [53].

Monolithic porous phosphorus containing organic porous polymers (PPOPs) with imine chain can be synthesized in one step by using Schiff base polycondensation reaction. They have regular three-dimensional network structure and good thermal insulation and flame retardancy [54].

5.3. *Schiff Base and Epoxy Resin*

Epoxy resin is a kind of key polymer material, which is used in coatings because of its excellent chemical stability and good mechanical properties. Epoxy resins are usually used for protection in corrosive environments. Their resistance to different corrosives depends on the type of resin, the type and amount of hardener, and the curing temperature [55]. However, the epoxy resin coating has high flammability, which will cause huge economic losses and personnel losses during use [56]. According to research, curing epoxy resin based on a Schiff base polymer has excellent adhesion to steel. For example, using protocatechol to prepare high-performance Schiff base epoxy resin, while maintaining excellent chemical resistance and thermal stability, gives it strong biological activity, degradability, excellent fire safety and antibacterial ability [53] (Figure 2).

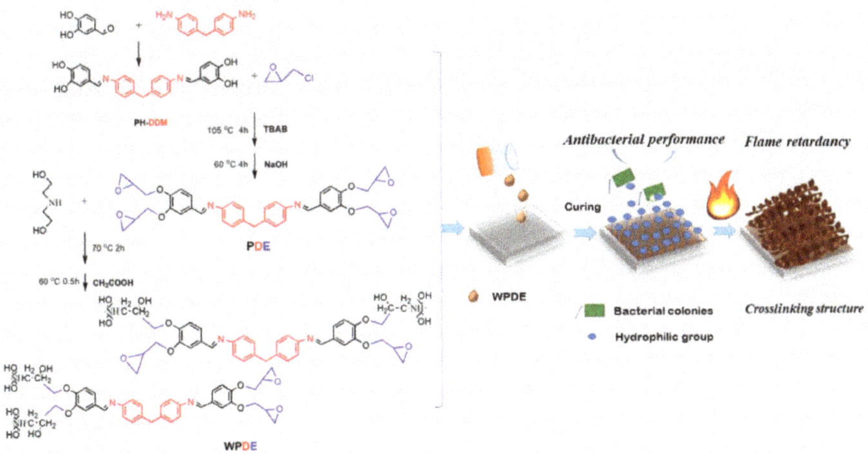

(a) (b)

Figure 2. (**a**) A synthetic route of Schiff base compounds (PH-DDM, PDE and WPDE) and (**b**) schematic diagram [53].

5.4. *Schiff Base and Others*

Structure determines performance. Based on its special structure, Schiff base has many unique properties. Schiff bases, as one of the most commonly used dynamic covalent bonds, have become an important research field due to their structural diversity and the simplicity of condensation of carbonyl compounds with primary amines. Moreover, Schiff bases can exchange with each other under external stimuli such as temperature and pH to trigger network topology rearrangement, thus endowing cross-linked polymers with many unique dynamic properties [57]. In addition, Schiff base metal complexes exhibit different biological and catalytic activities in different conversion processes [58], for example, a new Schiff base modified gold nano catalyst (its catalytic performance has a large size dependence) [16]. The Schiff base reaction is used to determine the binding sites of histones in nucleosomes and regulate RNA transcription [59]; It can also induce metal ions to achieve the detection purpose [60]. A new type of Schiff base ligand lanthanide complex was constructed by Schiff base reaction, which has low toxicity and high biological activity (anti-tumor, antibacterial, anti-HIV, anti-inflammatory and anti-cancer) [25].

Because most biomolecules contain amino groups in their structures, aldehyde-mediated Schiff base reactions have been widely used in the preparation of microarrays. The membrane that immobilizes the active enzyme structure on the solid surface through the interaction of Schiff bases can not only be used as a bioreactor, but is also often used as a biosensor for the determination of various biomolecules [15]. The transition of the unsaturated π bond in conjugation after Schiff base reaction leads to a certain self-fluorescence property of the product [61]. Schiff bases and their complexes are widely used as probes for ion detection in chemistry, environment, biology and other fields [18]. For example, the Schiff base exchange reaction is used to construct a fluorescent probe under the action of a catalyst. This probe is used for efficient and rapid detection of dichloromethane in aqueous solution [62]. At the same time, the donor receptor (D-A) and Schiff base coupling organic structure can also be used to reduce the infrared emissivity, thereby enhancing the infrared stealth performance of the polymer.

6. Schiff Base Coating

Based on the deep understanding of the molecular structure, preparation process and extensive application of Schiff base, Schiff base coatings have been widely used in the field of surface modification of biomedical materials. On the one hand, it can be used as a metal biomaterial coating to improve its molecular binding ability, and on the other hand, it can be used with other biomolecules to improve the biological function of the metal biomaterial surface.

6.1. Metal Coating

The Schiff base coating has excellent corrosion resistance [63]. Schiff base compounds are condensation products of amines and ketones/aldehydes. Recent publications show that these compounds are increasingly used as corrosion inhibitors for various metals (such as steel, aluminum and copper), especially in acidic environments [64]. Schiff base compounds are usually adsorbed on metal surfaces to block active corrosion sites to inhibit metal corrosion [65]. Schiff bases are introduced as a group of well-defined self-assembled monolayer-forming materials as corrosion inhibitors for steel, zinc and copper in different corrosion environments [66]. Moreover, some studies show that the inhibition efficiency of Schiff bases is much higher than that of corresponding aldehydes and amines [65]. A new type of Schiff base has been constructed to inhibit the corrosion of concrete reinforcement. The analysis of electrochemical and thermodynamic results shows that the protection process of this Schiff base inhibitor conforms to the Langmuir isotherm [67]. A "semi amphiphilic" copolymer/guanine conjugate was prepared by Schiff base reaction. Using the pH response characteristics of the dynamic ammonia bond, guanine, as a corrosion inhibitor, can be stably wrapped and continuously released when corrosion occurs [68]. A new type of Schiff base silicon oxide anti-corrosion coating was constructed, as an effective anti-corrosion barrier, which slowed down the corrosion and pitting potential of the alloy [69]. In addition, inspired by the natural modules of bacterial secretions, animal and plant extracts, Liu et al. synthesized Schiff compounds through tobramycin (TOB) of streptomyces and protocatechualdehyde (PR) of black fern. In addition, a dynamic self-renewing Schiff base metal composite coating (Fe/TOB-PR) n was prepared by a layer-by-layer self-assembly (LBL) method. It has been proved to be a universal coating that can be attached to different types of substrates, thus providing a research idea for the preparation of environment-friendly biological antifouling coatings [70].

Too fast degradation has always been the bottleneck of further application of magnesium (Mg) alloys in the biomedical field [71]. Although a large number of Mg alloy corrosion resistant coatings have been developed, there is not yet a coating that is completely suitable for Mg alloys (Figure 3) [72]. Recently, a Schiff base synthesized from amino acid and natural product paeonol was proved to have a good corrosion inhibition function for Mg alloy (Figure 4) [73], and was made into a coating for a Mg alloy substrate by electrostatic spraying (Figure 5) [74]. This Schiff base can chelate with the Mg ions degraded by Mg alloy and prevent the release of degradation products, thus the corrosion resistance of Mg alloy has been improved (Figure 6). This research also demonstrated that the coating made of the mixture of several single Schiff bases can better improve the corrosion resistance of Mg alloys, but its mechanism is still unclear.

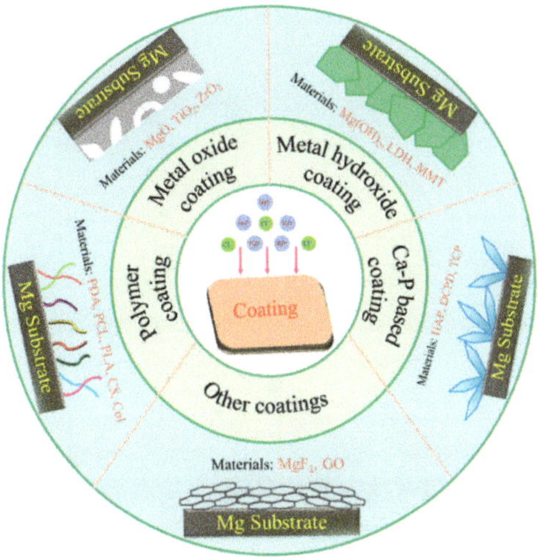

Figure 3. Classification of surface coatings according to their compositions [72].

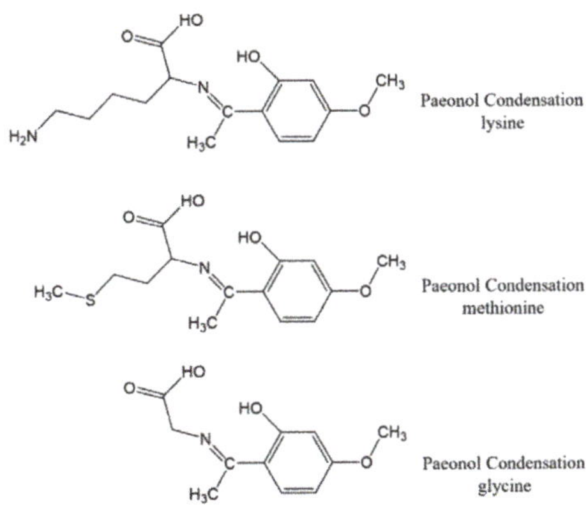

Figure 4. Molecular structures of three different kinds of Schiff base: PCLys, PCGly and PCMet [73].

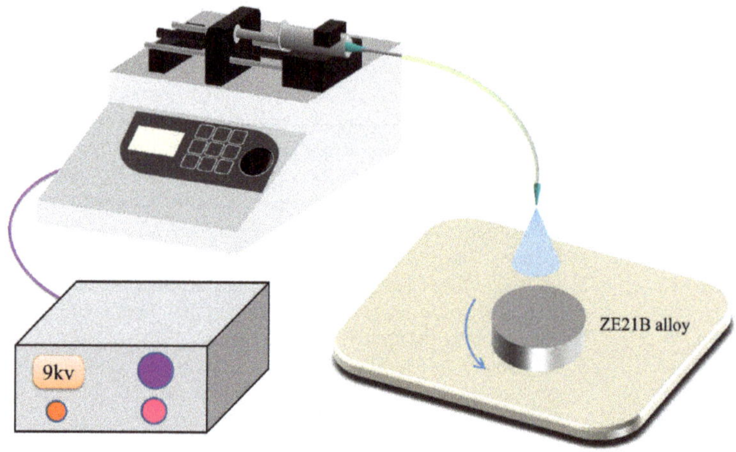

Figure 5. Schematic diagram of electrostatic spraying [74].

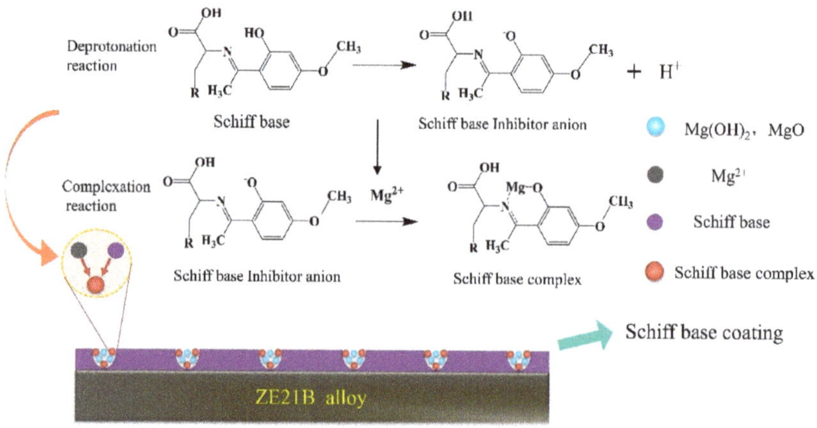

Figure 6. Schematic diagram of corrosion inhibition mechanism of Schiff base coating.

At present, biodegradable metals have been widely studied in the field of biomedical devices and have shown great application potential, such as iron-based cardiovascular stents [75], zinc-based bone implants [76], etc. Wherein, Mg alloy is one of the biodegradable metals with the fastest degradation rate, which has seriously affected its further clinical application [77]. For inhibiting the degradation rate of Mg alloy, many kinds of coatings have been reported at present, including magnesium fluoride (MgF_2) [78], magnesium hydroxide [79], etc. However, the preparation time of these coatings is too long. Mg alloy small devices, such as stents, can easily lose their precise structure or mechanical properties due to excessive corrosion during long-term acid or alkali passivation. In addition, it is reported that the Mg^{2+} produced by Mg alloy degradation can regulate the conversion of inflammatory M1 type macrophages to anti-inflammatory M2 type macrophages, and also promote the growth and function of vascular endothelial cells, but this function is limited by the concentration of Mg^{2+} released by Mg alloy [80]. This discovery proves that Mg alloy, as a vascular stent material, not only plays its mechanical properties and biodegradability, but the degradation products also can be used as an active factor in the treatment of diseases. Thus, the amino acid-paeonol Schiff base can be used as the exclusive coating of vascular stents due to its biological safety, sprayability and good corrosion resistance, so as to regulate the concentration of Mg^{2+} released by controlling degradation to meet the needs of various cells. Other advantages of Schiff base as the exclusive coating of Mg alloy for vascular stents are also summarized in Table 2.

Table 2. Properties comparison of corrosion resistance coatings on magnesium alloy surface.

Type	Schiff Base Coating	MgF_2	Magnesium Hydroxide
Corrosion resistance	+	+	+
Shorter preparation time	+	−	−
Better biocompatibility	+	−	−
Surface self-healing	+	−	−

"+" indicated possessing or better, "−" indicated not possessing or worse.

6.2. Biological Coating

Metal coatings also affect the biocompatibility of materials, so Schiff base coatings are also commonly used in implant materials. Surface modification of immobilized biomolecules has been widely demonstrated to enhance the biocompatibility of cardiovascular implants. Anti-CD133 and fucoidan were chemically fixed on the polydopamine (PDA) membrane known for its stability and endothelial cell (EC) compatibility (Michael addition and Schiff base reaction) to make a coating. In vitro experiments on New Zealand white rabbits showed that the coating had good blood compatibility and was expected to be applied to the surface modification of cardiovascular biomaterials [81]. Through the Schiff base reaction and Michael addition reaction, a functional coating with polymethylmethacrylate as the substrate was constructed, which reduced the protein surface adsorption capacity of the substrate and greatly improved the antifouling capacity [82]. Similarly, this reaction is used to prepare poly(vinylidene fluoride) (PVDF) membrane with underwater super oil repellency effect, which is used for environmental protection one-step modification research of efficient emulsion separation [83]. Schiff base prepared from glucose and amino acid is used to modify Ca-P coating. The modified Ca-P coating has fine particles, low surface roughness and high adhesion, which greatly improves its degradation in vitro [20]. Schiff base compounds were synthesized with p-phenylenediamine as crosslinking agent and reacted with polylactic acid (PLA) to reduce the brittleness of PLA biological composite coatings and improve their antibacterial activity [84]. Composite Schiff base coated samples have high resistance and good corrosion resistance in electrochemical polarization reaction. Schiff base coating can also promote endothelial cell growth and inhibit platelet adhesion/activation [74].

Poly-l-lysine-3,4-dihydroxybenzylaldehyde (PLL-DHBA) films can be prepared by cathodic electrophoretic deposition (EPD) based on the Schiff base reaction of amino groups of PLL monomers and aldehyde groups of 3,4-dihydroxybenzaldehyde (DHBA) molecules. Nanocomposite coatings with dual micro nano morphology have been developed for orthopedic and dental coatings. This work paved the way for the development of the next generation of biomedical implant coatings [34]. A new chitosan Schiff base and its Fe_2O_3 nanocomposite, as a natural adsorbent with low cost, good biocompatibility and biodegradability, have good performance in removing methyl orange [85]. It has a good prospect to be used on the surface of metal biomaterials. The new modified Schiff base nanoparticles obtained by crosslinking chitosan with bis (4-formyl-2-methoxyphenyl carbonate) have free radical scavenging ability and anti-cancer properties. The carbonate skeleton and part of the nanoparticles are hydrolyzed under the targeted carcinogenic microenvironment, releasing vanillin and chitosan, and enhancing the anti-cancer activity [32]. Chemical modification of chitosan with aromatic aldehydes such as vanillin, salicylaldehyde and provanillin can prepare stable Schiff base ligands [33].

Natural hydrogels have been widely studied in the field of biomedicine because of their good biocompatibility and biodegradability due to their structural similarity to the extracellular matrix of primary tissue. However, they are often vulnerable to mechanical failures. A new hydrogel was constructed by Schiff base reaction. Due to the existence of dynamic transfer base bonds, it has pH responsive swelling behavior and good mechanical properties, and has excellent self-healing and thixotropy [86]. In addition, it was found to be pH responsive and injectable. This hyaluronic acid gel has broad application prospects in drug release, biological printing, intelligent robots and tissue regeneration [87]. A kind of adhesive smart hydrogel inspired by double cross-linked mussel has good mechanical properties, stronger antibacterial and angiogenesis properties on the basis of maintaining the above properties, and is used for the treatment of chronic infected diabetes wounds [88]. Through the in situ formation reaction of Schiff base, hydrogels extracted from natural polysaccharides can be modulated and prepared for biomedical applications such as soft tissue adhesives and hemostatic agents in the future [51]. Nano-emulsion-loaded hydrogel coating is used to inhibit bacterial toxicity and the formation of solid surface biofilm [89]. The chitosan bound hydrogel has potential research value in the treatment of hepatocellular carcinoma [90].

A complex of Schiff base and copper metal was synthesized, and the modified coating was prepared by adding it to polyurethane varnish. It has good flame retardancy and antibacterial properties, and does not affect the flexibility, hardness and adhesion of polyurethane varnish [26].

7. Conclusions

I. Schiff bases are widely studied due to their synthetic flexibility, selectivity, sensitivity to central metal atoms, structural similarity with natural biological compounds, and their (− N-CH -) groups.

II. Schiff bases can be applied for surface modification of biomedical metals as functional coatings to improve the biocompatibility, comprehensive mechanical properties, biological activity and corrosion resistance.

III. An amino acid—paeonol Schiff base can endow the surface of Mg alloy with self-healing function and promote the growth of endothelial cells. Spraying this Schiff base on the surface in a few minutes is expected to make a special coating of Mg alloy for vascular stents.

IV. Many Schiff bases and their compound coatings still have unclear application mechanisms, which need further exploration.

Author Contributions: Conceptualization, J.L. and K.Z.; methodology, J.L.; investigation, Z.Z., Q.S. and Y.F.; writing—original draft preparation, Z.Z., Q.S. and Y.J.; writing—review and editing, J.L. and K.Z.; supervision, J.L.; funding acquisition, J.L., Y.F. and K.Z. All authors have read and agreed to the published version of the manuscript.

Funding: This research was funded by the National Natural Science Foundation of China, grant number U2004164; the Key Scientific and Technological Research Projects in Henan Province, grant number 222102310234 and 222102230025.

Data Availability Statement: Not applicable.

Acknowledgments: Table 1 reprinted (adapted) with permission from Ref. [18], Copyright {2022} American Chemical Society; Figure 2 reproduced from Ref. [53] with permission from Elsevier B.V. (License Number: 5452300869071).

Conflicts of Interest: The authors declare no conflict of interest.

References

1. Walter, T.; Gruenewald, A.; Detsch, R.; Boccaccini, A.R.; Vogel, N. Cell Interactions with Size-Controlled Colloidal Monolayers: Toward Improved Coatings in Bone Tissue Engineering. *Langmuir* **2020**, *36*, 1793–1803. [CrossRef] [PubMed]
2. Liu, X.; Sheng, S.; Yang, H.; He, Z.; Yang, Y.; Sheng, N.; Fang, H.; Shi, G. Uniform, Anticorrosive, and Antiabrasive Coatings on Metallic Surfaces for Cation-Metal and Cation-pi Interactions. *ACS Appl. Mater. Interfaces* **2020**, *12*, 38638–38646. [CrossRef] [PubMed]
3. Qi, Y.; Qi, H.; He, Y.; Lin, W.; Li, P.; Qin, L.; Hu, Y.; Chen, L.; Liu, Q.; Sun, H.; et al. Strategy of Metal-Polymer Composite Stent To Accelerate Biodegradation of Iron-Based Biomaterials. *ACS Appl. Mater. Interfaces* **2018**, *10*, 182–192. [CrossRef] [PubMed]
4. Soliman, A.I.A.; Sayed, M.; Elshanawany, M.M.; Younis, O.; Ahmed, M.; El-Dean, A.M.K.; Abdel-Wahab, A.M.A.; Wachtveitl, J.; Braun, M.; Fatehi, P.; et al. Base-Free Synthesis and Photophysical Properties of New Schiff Bases Containing Indole Moiety. *ACS Omega* **2022**, *7*, 10178–10186. [CrossRef]

5. Hassan, N.; Hendy, A.; Ebrahim, A.; Tamer, T.M. Synthesis and evaluation of novel O-functionalized aminated chitosan derivatives as antibacterial, antioxidant and anticorrosion for 316L stainless steel in simulated body fluid. *J. Saudi Chem. Soc.* **2021**, *25*, 101368. [CrossRef]
6. Liang, K.; Ricco, R.; Doherty, C.M.; Styles, M.J.; Bell, S.; Kirby, N.; Mudie, S.; Haylock, D.; Hill, A.J.; Doonan, C.J.; et al. Biomimetic mineralization of metal-organic frameworks as protective coatings for biomacromolecules. *Nat. Commun.* **2015**, *6*, 7240. [CrossRef]
7. Kindi, H.; Willems, C.; Zhao, M.; Menzel, M.; Schmelzer, C.E.H.; Herzberg, M.; Fuhrmann, B.; Gallego-Ferrer, G.; Groth, T. Metal Ion Doping of Alginate-Based Surface Coatings Induces Adipogenesis of Stem Cells. *ACS Biomater. Sci. Eng.* **2022**, *8*, 4327–4340. [CrossRef]
8. Wen, J.G.; Geng, W.; Geng, H.Z.; Zhao, H.; Jing, L.C.; Yuan, X.T.; Tian, Y.; Wang, T.; Ning, Y.J.; Wu, L. Improvement of Corrosion Resistance of Waterborne Polyurethane Coatings by Covalent and Noncovalent Grafted Graphene Oxide Nanosheets. *ACS Omega* **2019**, *4*, 20265–20274. [CrossRef]
9. Fadl, A.M.; Abdou, M.I.; Laila, D.; Sadeek, S.A. Application insights of Schiff base metal complex/SiO_2 hybrid epoxy nanocomposite for steel surface coating: Correlation the protective behavior and mechanical properties with material loading. *Prog. Org. Coat.* **2019**, *136*, 105226. [CrossRef]
10. El-Azabawy, O.E.; Higazy, S.A.; Al-Sabagh, A.M.; Abdel-Rahman, A.A.H.; Nasser, N.M.; Khamis, E.A. Studying the temperature influence on carbon steel in sour petroleum media using facilely-designed Schiff base polymers as corrosion inhibitors. *J. Mol. Struct.* **2023**, *1275*, 134518. [CrossRef]
11. Long, W.-J.; Li, X.-Q.; Xu, P.; Feng, G.-L.; He, C. Facile and scalable preparation of carbon dots with Schiff base structures toward an efficient corrosion inhibitor. *Diam. Relat. Mater.* **2022**, *130*, 109401. [CrossRef]
12. Jafari, H.; Ameri, E.; Rezaeivala, M.; Berisha, A. Experimental and theoretical studies on protecting steel against 0.5 M H_2SO_4 corrosion by new schiff base. *J. Indian Chem. Soc.* **2022**, *99*, 100665. [CrossRef]
13. El-Ghamry, M.A.; Elzawawi, F.M.; Aziz, A.A.A.; Nassir, K.M.; Abu-El-Wafa, S.M. New Schiff base ligand and its novel Cr(III), Mn(II), Co(II), Ni(II), Cu(II), Zn(II) complexes: Spectral investigation, biological applications, and semiconducting properties. *Sci. Rep.* **2022**, *12*, 17942. [CrossRef] [PubMed]
14. Zhang, W.; Shi, T.; Ding, G.; Punyapitak, D.; Zhu, J.; Guo, D.; Zhang, Z.; Li, J.; Cao, Y. Nanosilica Schiff-Base Copper(II) Complexes with Sustainable Antimicrobial Activity against Bacteria and Reduced Risk of Harm to Plant and Environment. *ACS Sustain. Chem. Eng.* **2016**, *5*, 502–509. [CrossRef]
15. Jia, Y.; Li, J. Molecular assembly of Schiff Base interactions: Construction and application. *Chem. Rev.* **2015**, *115*, 1597–1621. [CrossRef]
16. Liu, Q.; Yang, X.; Li, L.; Miao, S.; Li, Y.; Li, Y.; Wang, X.; Huang, Y.; Zhang, T. Direct catalytic hydrogenation of $CO_{(2)}$ to formate over a Schiff-base-mediated gold nanocatalyst. *Nat. Commun.* **2017**, *8*, 1407. [CrossRef]
17. Luo, C.; Li, M.; Yuan, R.; Yang, Y.; Lu, Z.; Ge, L. Biocompatible Self-Healing Coating Based on Schiff Base for Promoting Adhesion of Coral Cells. *ACS Appl. Bio Mater.* **2020**, *3*, 1481–1495. [CrossRef]
18. Zhong, X.; Li, Z.; Shi, R.; Yan, L.; Zhu, Y.; Li, H. Schiff Base-Modified Nanomaterials for Ion Detection: A Review. *ACS Appl. Nano Mater.* **2022**, *5*, 13998–14020. [CrossRef]
19. Negm, N.A.; Altalhi, A.A.; Mohamed, N.E.S.; Kana, M.T.H.A.; Mohamed, E.A. Growth Inhibition of Sulfate-Reducing Bacteria during Gas and Oil Production Using Novel Schiff Base Diquaternary Biocides: Synthesis, Antimicrobial, and Toxicological Assessment. *ACS Omega* **2022**, *7*, 40098–40108. [CrossRef]
20. Wang, X.; Fan, X.; Zeng, M.; Li, C.; Cui, L.; Chen, X.; Zou, Y.; Wang, Z.; Zeng, R. In vitro degradation resistance of glucose and L-cysteine-bioinspired Schiff-base anodic Ca–P coating on AZ31 magnesium alloy. *Trans. Nonferr. Met. Soc. China* **2022**, *32*, 1485–1500. [CrossRef]
21. Verma, R.; Lamba, N.P.; Dandia, A.; Srivastava, A.; Modi, K.; Chauhan, M.S.; Prasad, J. Synthesis of N-Benzylideneaniline by Schiff base reaction using Kinnow peel powder as Green catalyst and comparative study of derivatives through ANOVA techniques. *Sci. Rep.* **2022**, *12*, 9636. [CrossRef] [PubMed]
22. Beyazit, N.; Çakran, H.S.; Cabir, A.; Akışcan, Y.; Demetgül, C. Synthesis, characterization and antioxidant activity of chitosan Schiff base derivatives bearing (-)-gossypol. *Carbohydr. Polym.* **2020**, *240*, 116333. [CrossRef] [PubMed]
23. Zhang, H.; Liu, P.; Peng, X.; Chen, S.; Zhang, K. Interfacial Synthesis of Cellulose-Derived Solvent-Responsive Nanoparticles via Schiff Base Reaction. *ACS Sustain. Chem. Eng.* **2019**, *7*, 16595–16603. [CrossRef]
24. Tachibana, Y.; Hayashi, S.; Kasuya, K.I. Biobased Poly(Schiff-Base) Composed of Bifurfural. *ACS Omega* **2018**, *3*, 5336–5345. [CrossRef]
25. Andiappan, K.; Sanmugam, A.; Deivanayagam, E.; Karuppasamy, K.; Kim, H.S.; Vikraman, D. In vitro cytotoxicity activity of novel Schiff base ligand-lanthanide complexes. *Sci. Rep.* **2018**, *8*, 3054. [CrossRef]
26. Abd El-Wahab, H.; El-Fattah, M.A.; El-alfy, H.M.Z.; Owda, M.E.; Lin, L.; Hamdy, I. Synthesis and characterisation of sulphonamide (Schiff base) ligand and its copper metal complex and their efficiency in polyurethane varnish as flame retardant and antimicrobial surface coating additives. *Prog. Org. Coat.* **2020**, *142*, 105577. [CrossRef]
27. Wang, J.; Meng, Q.; Yang, Y.; Zhong, S.; Zhang, R.; Fang, Y.; Gao, Y.; Cui, X. Schiff Base Aggregation-Induced Emission Luminogens for Sensing Applications: A Review. *ACS Sens.* **2022**, *7*, 2521–2536. [CrossRef]
28. Vikneshvaran, S.; Velmathi, S. Adsorption of L-Tryptophan-derived chiral Schiff bases on stainless steel surface for the prevention of corrosion in acidic environment: Experimental, theoretical and surface studies. *Surf. Interfaces* **2017**, *6*, 134–142. [CrossRef]
29. Hou, C.-C.; Ma, C.; Zhang, S.-N.; Wang, L.-Y.; Wang, K.-X.; Chen, J.-S. Polymeric Schiff Base with Thiophene Rings for Sodium-Ion Batteries. *ACS Appl. Energy Mater.* **2022**, *5*, 13802–13807. [CrossRef]

30. Marvaniya, K.; Maurya, A.; Dobariya, P.; Kaushik, A.; Prakash, P.; Bhargava, J.; Vanamudan, A.; Patel, K.; Kushwaha, S. Polymeric nano-films with spatially arranged compartments for uranium recovery from seawater. *Eur. Polym. J.* **2022**, *178*, 111507. [CrossRef]
31. Ani, F.E.; Ibeji, C.U.; Obasi, N.L.; Kelani, M.T.; Ukogu, K.; Tolufashe, G.F.; Ogundare, S.A.; Oyeneyin, O.E.; Maguire, G.E.M.; Kruger, H.G. Crystal, spectroscopic and quantum mechanics studies of Schiff bases derived from 4-nitrocinnamaldehyde. *Sci. Rep.* **2021**, *11*, 8151. [CrossRef] [PubMed]
32. Thyriyalakshmi, P.; Radha, K.V. Fabrication of chitosan-bis (4-formyl-2 methoxy phenyl carbonate) Schiff base nanoparticles and evaluation of their antioxidant and anticancer properties. *Mol. Biol. Rep.* **2019**, *46*, 4333–4347. [CrossRef] [PubMed]
33. Vadivel, T.; Dhamodaran, M.; Kulathooran, S.; Kavitha, S.; Amirthaganesan, K.; Chandrasekaran, S.; Ilayaraja, S.; Senguttuvan, S. Rhodium(III) complexes derived from complexation of metal with azomethine linkage of chitosan biopolymer Schiff base ligand: Spectral, thermal, morphological and electrochemical studies. *Carbohydr. Res.* **2020**, *487*, 107878. [CrossRef] [PubMed]
34. Clifford, A.; Lee, B.E.J.; Grandfield, K.; Zhitomirsky, I. Biomimetic modification of poly-l-lysine and electrodeposition of nanocomposite coatings for orthopaedic applications. *Colloids Surf. B Biointerfaces* **2019**, *176*, 115–121. [CrossRef] [PubMed]
35. Jehle, F.; Macías-Sánchez, E.; Sviben, S.; Fratzl, P.; Bertinetti, L.; Harrington, M.J. Hierarchically-structured metalloprotein composite coatings biofabricated from co-existing condensed liquid phases. *Nat. Commun.* **2020**, *11*, 862. [CrossRef] [PubMed]
36. Elshaarawy, R.F.M.; Mustafa, F.H.A.; van Geelen, L.; Abou-Taleb, A.E.A.; Tadros, H.R.Z.; Kalscheuer, R.; Janiak, C. Mining marine shell wastes for polyelectrolyte chitosan anti-biofoulants: Fabrication of high-performance economic and ecofriendly anti-biofouling coatings. *Carbohydr. Polym.* **2017**, *172*, 352–364. [CrossRef]
37. Hassan, M.A.; Omer, A.M.; Abbas, E.; Baset, W.M.A.; Tamer, T.M. Preparation, physicochemical characterization and antimicrobial activities of novel two phenolic chitosan Schiff base derivatives. *Sci. Rep.* **2018**, *8*, 11416. [CrossRef]
38. Han, J.; Wang, H.; Li, Z.; Wang, Z. Preparation of chitosan-modified magnetic Schiff base network composite nanospheres for effective enrichment and detection of hippuric acid and 4-methyl hippuric acid. *J. Chromatogr. A* **2021**, *1652*, 462373. [CrossRef]
39. Manimohan, M.; Paulpandiyan, R.; Pugalmani, S.; Sithique, M.A. Biologically active Co (II), Cu (II), Zn (II) centered water soluble novel isoniazid grafted O-carboxymethyl chitosan Schiff base ligand metal complexes: Synthesis, spectral characterisation and DNA nuclease activity. *Int. J. Biol. Macromol.* **2020**, *163*, 801–816. [CrossRef]
40. Ali, S.S.; Kenawy, E.R.; Sonbol, F.I.; Sun, J.; Al-Etewy, M.; Ali, A.; Liu, H.; El-Zawawy, N.A. Pharmaceutical Potential of a Novel Chitosan Derivative Schiff Base with Special Reference to Antibacterial, Anti-Biofilm, Antioxidant, Anti-Inflammatory, Hemocompatibility and Cytotoxic Activities. *Pharm. Res.* **2018**, *36*, 5. [CrossRef]
41. Malekshah, R.E.; Shakeri, F.; Khaleghian, A.; Salehi, M. Developing a biopolymeric chitosan supported Schiff-base and Cu(II), Ni(II) and Zn(II) complexes and biological evaluation as pro-drug. *Int. J. Biol. Macromol.* **2020**, *152*, 846–861. [CrossRef] [PubMed]
42. Khan, A.M.; Abid, O.U.R.; Mir, S. Assessment of biological activities of chitosan Schiff base tagged with medicinal plants. *Biopolymers* **2020**, *111*, e23338. [CrossRef] [PubMed]
43. Anush, S.M.; Chandan, H.R.; Vishalakshi, B. Synthesis and metal ion adsorption characteristics of graphene oxide incorporated chitosan Schiff base. *Int. J. Biol. Macromol.* **2019**, *126*, 908–916. [CrossRef] [PubMed]
44. Miao, Q.; Mi, Y.; Cui, J.; Zhang, J.; Tan, W.; Li, Q.; Guo, Z. Determination of chitosan content with Schiff base method and HPLC. *Int. J. Biol. Macromol.* **2021**, *182*, 1537–1542. [CrossRef] [PubMed]
45. Ansari, K.R.; Chauhan, D.S.; Quraishi, M.A.; Mazumder, M.A.J.; Singh, A. Chitosan Schiff base: An environmentally benign biological macromolecule as a new corrosion inhibitor for oil & gas industries. *Int. J. Biol. Macromol.* **2020**, *144*, 305–315.
46. Seidi, S.; Majd, M.; Rezazadeh, M.; Shanehsaz, M. Magnetic nanocomposite of chitosan-Schiff base grafted graphene oxide for lead analysis in whole blood. *Anal. Biochem.* **2018**, *553*, 28–37. [CrossRef]
47. Rakhtshah, J.; Yaghoobi, F. Catalytic application of new manganese Schiff-base complex immobilized on chitosan-coated magnetic nanoparticles for one-pot synthesis of 3-iminoaryl-imidazo[1,2-a]pyridines. *Int. J. Biol. Macromol.* **2019**, *139*, 904–916. [CrossRef]
48. Arnon-Rips, H.; Cohen, Y.; Saidi, L.; Porat, R.; Poverenov, E. Covalent linkage of bioactive volatiles to a polysaccharide support as a potential approach for preparing active edible coatings and delivery systems for food products. *Food Chem.* **2021**, *338*, 127822. [CrossRef]
49. Xu, Y.; Shi, Y.; Lei, F.; Dai, L. A novel and green cellulose-based Schiff base-Cu (II) complex and its excellent antibacterial activity. *Carbohydr. Polym.* **2020**, *230*, 115671. [CrossRef]
50. Zhong, W.; Xiong, Y.; Wang, X.; Yu, T.; Zhou, C. Synthesis and characterization of multifunctional organic-inorganic composite hydrogel formed with tissue-adhesive property and inhibiting infection. *Mater. Sci. Eng. C Mater. Biol. Appl.* **2021**, *118*, 111532. [CrossRef]
51. Liu, J.; Li, J.; Yu, F.; Zhao, Y.X.; Mo, X.M.; Pan, J.F. In situ forming hydrogel of natural polysaccharides through Schiff base reaction for soft tissue adhesive and hemostasis. *Int. J. Biol. Macromol.* **2020**, *147*, 653–666. [CrossRef]
52. Deng, H.; Huang, K.; Xiu, L.; Sun, W.; Yao, Q.; Fang, X.; Huang, X.; Noreldeen, H.A.A.; Peng, H.; Xie, J.; et al. Bis-Schiff base linkage-triggered highly bright luminescence of gold nanoclusters in aqueous solution at the single-cluster level. *Nat. Commun.* **2022**, *13*, 3381. [CrossRef]
53. Ji, J.; Huang, S.; Liu, S.; Yuan, Y.; Zhao, J.; Zhang, S. A novel biomass-derived Schiff base waterborne epoxy coating for flame retardation and anti-bacteria. *Polym. Degrad. Stab.* **2022**, *199*, 109910. [CrossRef]
54. Zhao, P.; Tian, L.; Guo, Y.; Lv, B.; Mao, X.; Li, T.; Cui, J.; Guo, J.; Yang, B. A facile method to prepare high-performance thermal insulation and flame retardant materials from amine-linked porous organic polymers. *Eur. Polym. J.* **2022**, *162*, 110918. [CrossRef]

55. Atta, A.M.; Shaker, N.O.; Maysour, N.E. Influence of the molecular structure on the chemical resistivity and thermal stability of cured Schiff base epoxy resins. *Prog. Org. Coat.* **2006**, *56*, 100–110. [CrossRef]
56. Ma, J.; Li, G.; Hua, X.; Liu, N.; Liu, Z.; Zhang, F.; Yu, L.; Chen, X.; Shang, L.; Ao, Y. Biodegradable epoxy resin from vanillin with excellent flame-retardant and outstanding mechanical properties. *Polym. Degrad. Stab.* **2022**, *201*, 109989. [CrossRef]
57. Li, J.; Weng, Z.; Cao, Q.; Qi, Y.; Lu, B.; Zhang, S.; Wang, J.; Jian, X. Synthesis of an aromatic amine derived from biomass and its use as a feedstock for versatile epoxy thermoset. *Chem. Eng. J.* **2022**, *433*, 134512. [CrossRef]
58. Daraie, M.; Heravi, M.M.; Rangraz, Y.; Besharati, Z. Pd NPs supported on halloysite functionalized with Schiff base as an efficient catalyst for Sonogashira reaction. *Sci. Rep.* **2021**, *11*, 6223. [CrossRef]
59. Raiber, E.A.; Portella, G.; Cuesta, S.M.; Hardisty, R.; Murat, P.; Li, Z.; Iurlaro, M.; Dean, W.; Spindel, J.; Beraldi, D.; et al. 5-Formylcytosine organizes nucleosomes and forms Schiff base interactions with histones in mouse embryonic stem cells. *Nat. Chem.* **2018**, *10*, 1258–1266. [CrossRef]
60. Muthusamy, S.; Zhu, D.; Rajalakshmi, K.; Zhu, W.; Wang, S.; Lee, K.B.; Zhao, L. Successive Detection of Zinc Ion and Citrate Using a Schiff Base Chemosensor for Enhanced Prostate Cancer Diagnosis in Biosystems. *ACS Appl. Bio Mater.* **2021**, *4*, 1932–1941. [CrossRef]
61. Ge, L.; Li, Z.; Han, M.; Wang, Y.; Li, X.; Mu, C.; Li, D. Antibacterial dialdehyde sodium alginate/epsilon-polylysine microspheres for fruit preservation. *Food Chem.* **2022**, *387*, 132885. [CrossRef]
62. Mo, K.Q.; Ma, X.F.; Wang, H.L.; Zhu, Z.H.; Liu, Y.C.; Zou, H.H.; Liang, F.P. Tracking the Multistep Formation of Ln(III) Complexes with in situ Schiff Base Exchange Reaction and its Highly Selective Sensing of Dichloromethane. *Sci. Rep.* **2019**, *9*, 12231. [CrossRef]
63. Satpati, S.; Suhasaria, A.; Ghosal, S.; Adhikari, U.; Banerjee, P.; Dey, S.; Sukul, D. Anti-corrosive propensity of naturally occurring aldehydes and 1-(3-aminopropyl)imidazole condensed Schiff bases: Comparison on the effect of extended conjugation over electron donating substituents. *J. Mol. Struct.* **2022**, *1268*, 133684. [CrossRef]
64. Emregül, K.C.; Düzgün, E.; Atakol, O. The application of some polydentate Schiff base compounds containing aminic nitrogens as corrosion inhibitors for mild steel in acidic media. *Corros. Sci.* **2006**, *48*, 3243–3260. [CrossRef]
65. Prabhu, R.A.; Venkatesha, T.V.; Shanbhag, A.V.; Kulkarni, G.M.; Kalkhambkar, G.R. Inhibition effects of some Schiff's bases on the corrosion of mild steel in hydrochloric acid solution. *Corros. Sci.* **2008**, *50*, 3356–3362. [CrossRef]
66. Ehteshamzade, M.; Shahrabi, T.; Hosseini, M.G. Inhibition of copper corrosion by self-assembled films of new Schiff bases and their modification with alkanethiols in aqueous medium. *Appl. Surf. Sci.* **2006**, *252*, 2949–2959. [CrossRef]
67. Bellal, Y.; Benghanem, F.; Keraghel, S. A new corrosion inhibitor for steel rebar in concrete: Synthesis, electrochemical and theoretical studies. *J. Mol. Struct.* **2021**, *1225*, 129257. [CrossRef]
68. Cao, Y.; He, J.; Wu, J.; Wang, X.; Lu, W.; Lin, J.; Xu, Y.; Chen, G.; Zeng, B.; Dai, L. A Smart Anticorrosive Epoxy Coating Based on Environmental-Stimuli-Responsive Copolymer Assemblies for Controlled Release of Corrosion Inhibitors. *Macromol. Mater. Eng.* **2022**, *307*, 2100983. [CrossRef]
69. Bonetti, S.; Spengler, R.; Petersen, A.; Aleixo, L.S.; Merlo, A.A.; Tamborim, S.M. Surface-decorated silica with Schiff base as an anticorrosive coating for aluminium alloy 2024-T3. *Appl. Surf. Sci.* **2019**, *475*, 684–694. [CrossRef]
70. Liu, W.; Yan, M.; Zhao, W. Antibacterial-renew dual-function anti-biofouling strategy: Self-assembled Schiff-base metal complex coatings built from natural products. *J. Colloid Interface Sci.* **2023**, *629*(Pt. A), 496–507. [CrossRef]
71. Li, J.; Zhou, P.; Wang, L.; Hou, Y.; Zhang, X.; Zhu, S.; Guan, S. Investigation of Mg-xLi-Zn alloys for potential application of biodegradable bone implant materials. *J. Mater. Sci. Mater. Med.* **2021**, *32*, 43. [CrossRef] [PubMed]
72. Tong, P.; Sheng, Y.; Hou, R.; Iqbal, M.; Chen, L.; Li, J. Recent progress on coatings of biomedical magnesium alloy. *Smart Mater. Med.* **2022**, *3*, 104–116. [CrossRef]
73. Sheng, Y.; Hou, R.; Liu, C.; Xue, Z.; Zhang, K.; Li, J.; Guan, S. Tailoring of biodegradable magnesium alloy surface with Schiff base coating via electrostatic spraying for better corrosion resistance. *Metals* **2022**, *12*, 471. [CrossRef]
74. Sheng, Y.; Li, W.; Chai, Y.; Yin, S.; Li, J.; Guan, S. A compound Schiff base coating on biomedical magnesium alloy for enhanced corrosion resistance and biocompatibility. *Smart Mater. Manuf.* **2023**, *1*, 100003. [CrossRef]
75. Lin, W.; Zhang, H.; Zhang, W.; Qi, H.; Zhang, G.; Qian, J.; Li, X.; Qin, L.; Li, H.; Wang, X.; et al. In vivo degradation and endothelialization of an iron bioresorbable scaffold. *Bioact. Mater.* **2021**, *6*, 1028–1039. [CrossRef]
76. Wu, H.; Xie, X.; Wang, J.; Ke, G.; Huang, H.; Liao, Y.; Kong, Q. Biological properties of Zn–0.04Mg–2Ag: A new degradable zinc alloy scaffold for repairing large-scale bone defects. *J. Mater. Res. Technol.* **2021**, *13*, 1779–1789. [CrossRef]
77. Gao, Y.; Zhao, P.; Cao, X.Q.; Misra, R.D.K.; Wang, W.; Chen, K.Z. Tin-induced microstructural changes and associated corrosion resistance of biodegradable Mg–Zn alloy. *Rare Met.* **2022**, *41*, 883–888. [CrossRef]
78. Hou, Y.; Zhang, X.; Li, J.; Wang, L.; Guan, S. A multi-functional MgF2/Polydopamine/Hyaluronan-astaxanthin coating on the biodegradable ZE21B alloy with better corrosion resistance and biocompatibility for cardiovascular application. *J. Magnes. Alloy.* 2022, *in press*. [CrossRef]
79. Han, Z.; Guo, H.; Zhou, Y.; Wang, L.; Zhang, K.; Li, J. Composite coating prepared with ferulic acid to improve the corrosion resistance and blood compatibility of magnesium alloy. *Metals* **2022**, *12*, 545. [CrossRef]
80. Hou, Y.; Witte, F.; Li, J.; Guan, S. The increased ratio of Mg^{2+}/Ca^{2+} from degrading magnesium alloys directs macrophage fate for functionalized growth of endothelial cells. *Smart Mater. Med.* **2022**, *3*, 188–198. [CrossRef]
81. Su, H.; Xue, G.; Ye, C.; Wang, Y.; Zhao, A.; Huang, N.; Li, J. The effect of anti-CD133/fucoidan bio-coatings on hemocompatibility and EPC capture. *J. Biomater. Sci. Polym. Ed.* **2017**, *28*, 2066–2081. [CrossRef] [PubMed]

82. Wen, H.; Hsu, Y.I.; Asoh, T.A.; Uyama, H. Poly(methyl methacrylate) surface grafted with poly(2-ethyl-2-oxazoline) using tea polyphenol as linker molecule. *Prog. Org. Coat.* **2022**, *166*, 106796. [CrossRef]
83. Zhu, T.; Jiang, C.; Wu, J.; Wang, M.; Zhu, C.; Zhao, N.; Xu, J. Eco-friendly and one-step modification of poly(vinylidene fluoride) membrane with underwater superoleophobicity for effective emulsion separation. *Colloids Surf. A Physicochem. Eng. Asp.* **2021**, *610*, 125939. [CrossRef]
84. Jain-Beuguel, C.; Li, X.; Houel-Renault, L.; Modjinou, T.; Simon-Colin, C.; Gref, R.; Renard, E.; Langloi, V. Water-Soluble Poly(3-hydroxyalkanoate) Sulfonate: Versatile Biomaterials Used as Coatings for Highly Porous Nano-Metal Organic Framework. *Biomacromolecules* **2019**, *20*, 3324–3332. [CrossRef] [PubMed]
85. Foroughnia, A.; Khalaji, A.D.; Kolvari, E.; Koukabi, N. Synthesis of new chitosan Schiff base and its Fe_2O_3 nanocomposite: Evaluation of methyl orange removal and antibacterial activity. *Int. J. Biol. Macromol.* **2021**, *177*, 83–91. [CrossRef]
86. Pandit, A.H.; Nisar, S.; Imtiyaz, K.; Nadeem, M.; Mazumdar, N.; Rizvi, M.M.A.; Ahmad, S. Injectable, Self-Healing, and Biocompatible N,O-Carboxymethyl Chitosan/Multialdehyde Guar Gum Hydrogels for Sustained Anticancer Drug Delivery. *Biomacromolecules* **2021**, *22*, 3731–3745. [CrossRef] [PubMed]
87. Li, S.; Pei, M.; Wan, T.; Yang, H.; Gu, S.; Tao, Y.; Liu, X.; Zhou, Y.; Xu, W.; Xiao, P. Self-healing hyaluronic acid hydrogels based on dynamic Schiff base linkages as biomaterials. *Carbohydr. Polym.* **2020**, *250*, 116922. [CrossRef]
88. Hu, C.; Long, L.; Cao, J.; Zhang, S.; Wang, Y. Dual-crosslinked mussel-inspired smart hydrogels with enhanced antibacterial and angiogenic properties for chronic infected diabetic wound treatment via pH-responsive quick cargo release. *Chem. Eng. J.* **2021**, *411*, 128564. [CrossRef]
89. Barik, P.S.K.; Singh, B.N. Nanoemulsion-loaded hydrogel coatings for inhibition of bacterial virulence and biofilm formation on solid surfaces. *Sci. Rep.* **2019**, *9*, 6520.
90. Elhag, M.; Abdelwahab, H.E.; Mostafa, M.A.; Yacout, G.A.; Nasr, A.Z.; Dambruoso, P.; Sadek, M.M.E. One pot synthesis of new cross-linked chitosan-Schiff′ base: Characterization, and anti-proliferative activities. *Int. J. Biol. Macromol.* **2021**, *184*, 558–565. [CrossRef]

metals

Review

Bioinspired Surface Design for Magnesium Alloys with Corrosion Resistance

Feng Wu [1,*], Yixuan Liu [1], Jing Xu [2] and Changjiang Pan [3,*]

1 College of Physics and New Energy, Xuzhou University of Technology, Xuzhou 221018, China
2 Medical Laboratory Department, The First People's Hospital of Xuzhou, Xuzhou 221116, China
3 Faculty of Mechanical and Material Engineering, Huaiyin Institute of Technology, Huai'an 223003, China
* Correspondence: wuf@xzit.edu.cn (F.W.); panchangjiang@hyit.edu.cn (C.P.)

Abstract: Magnesium alloys are regarded as potential candidates in industrial and biomedical applications because of their excellent mechanical properties and biodegradability. However, the excessive degradation rate of magnesium alloys can cause a premature disintegration of mechanical integrity, which is the main bottleneck that limits applications. Inspired by nature, various novel surface designs provide a clever strategy to regulate the corrosion behavior of magnesium alloys. This review extensively discusses bioinspired surface designs to reduce corrosion resistance and realize functionalization, so as to offer new ideas with great potential for biomedical applications. Future research on corrosion resistance is expected to benefit greatly from the bioinspired surface designs.

Keywords: bioinspired; surface design; magnesium alloys; corrosion

check for updates

Citation: Wu, F.; Liu, Y.; Xu, J.; Pan, C. Bioinspired Surface Design for Magnesium Alloys with Corrosion Resistance. *Metals* **2022**, *12*, 1404. https://doi.org/10.3390/met12091404

Academic Editor: Manuel Aureliano

Received: 20 July 2022
Accepted: 20 August 2022
Published: 24 August 2022

1. Introduction

Today, more than 200 years after the first isolation of elemental magnesium, biodegradable magnesium-based metals have attracted extensive attention [1–4]. The mechanical performances of magnesium alloys, such as plasticity, stiffness, and processability, are better than those of the bioresorbable polymers, such as polylactic acid employed in several clinical applications, due to the peculiarities of the metal properties [5,6]. Additionally, the density of magnesium alloys ranges from 1.73 to 1.85 g cm^{-3} and is similar to cortical bone (1.75 g cm^{-3}) [7,8]. As a result, it is anticipated that magnesium alloys may replace other materials [9–12].

Since the most negative self-corrosion potential is so low (−2.37 V), magnesium alloys can naturally dissolve and adsorb after being implanted in the body [13,14]. After two weeks, bladder, liver, and renal tissues show no adverse effects from the generated Mg^{2+}, which is mostly used by the organism [15–17]. However, the fundamental obstacle preventing more widespread use in industrial and medical settings is the rapid rate of deterioration of magnesium alloys. When magnesium alloys are exposed in a physiological environment, it is primarily degraded by electrochemical reactions, which produce magnesium hydroxide, hydrogen, and other compounds [18–20]. The corrosion of magnesium alloys also worsens due to the adsorption of biomolecules, including protein, small-molecule amino acids, and lipids [21,22]. Hydrogen gas produced by corrosion results in a somewhat alkaline atmosphere [23]. During the reaction process, a layer of magnesium hydroxide can also be produced to form the magnesium alloys' surface coat, the structure of which is flimsy and porous [24,25]. The corrosion medium can be further etched via these openings to create a new magnesium matrix, creating a pit that is corrosive and releases a lot of hydrogen gas [26].

Nature has supplied us with a multitude of inspirations [27,28] to overcome the difficulties in the design of protective surfaces for magnesium alloys. The bioinspired surface design used in this review is primarily focused on enhancing the corrosion resistance of magnesium alloys (Figure 1). An introduction to the corrosion mechanism of magnesium alloys is presented in the first part. The methods for creating bioinspired solid surface designs and liquid surface designs are shown in the next section. The conclusions are then offered, along with a brief prognosis of the development's course.

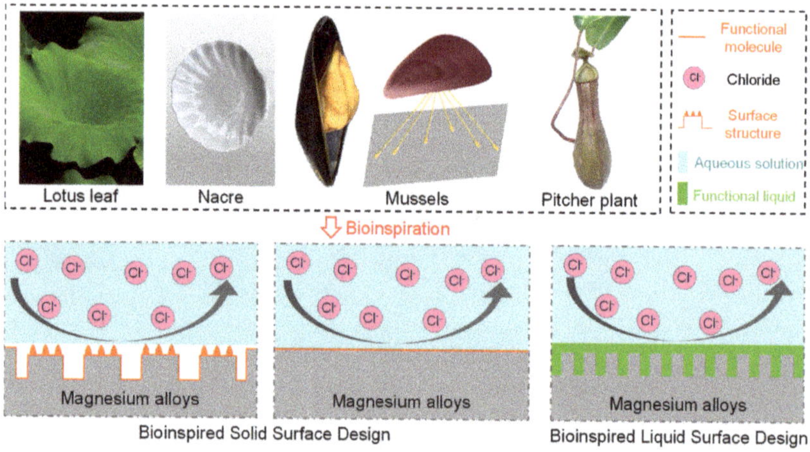

Figure 1. Schematic illustration of bioinspired surface designs to enhance the anticorrosion property.

2. Corrosion Mechanism

According to thermodynamics, the system's free energy occurs throughout the corrosion process [29,30]:

$$\Delta G = -nFE_0 \tag{1}$$

where ΔG denotes the change in the interfacial Gibbs free energy, E_0 the cell's standard potential, F the Faraday's constant, and n the quantity of electrons transferred during the process. The system experiences spontaneous oxidation when the Gibbs free energy is negative. Magnesium alloys quickly oxidize when in contact with water and liberate hydrogen gas from the water because of their weak thermodynamic stability [31]. Following are expressions for the corrosion process of magnesium alloys [32–34]:

$$2Mg \rightarrow Mg^{+} + 2e^{-}, \tag{2}$$

$$2Mg^{+} + 2H_2O \rightarrow 2Mg^{2+} + 2OH^{-} + H_2, \tag{3}$$

$$2H_2O + 2e^{-} \rightarrow 2OH^{-} + H_2, \tag{4}$$

$$2Mg^{2+} + 4OH^{-} \rightarrow 2Mg(OH)_2, \tag{5}$$

$$2MgO + H_2O \rightarrow Mg(OH)_2, \tag{6}$$

Equations (2)–(6) show that preventing magnesium alloys from contacting water helps to inhibit corrosion. Therefore, building a physical barrier on the surface of magnesium alloys exhibits great corrosion protection potential for magnesium alloys [35,36]. In comparison, bare magnesium alloys without a physical barrier are completely exposed to the corrosive solution. It is generally believed that boosting the compactness, stability, and thickness of the physical barrier surface can be an efficient strategy to increase the magnesium alloys' anticorrosion property from the standpoint of surface design.

Bioinspired surface design is a useful strategy for improving magnesium alloys' anticorrosion property by forming a protective layer to keep magnesium alloys from coming into direct touch with external aqueous solution, preventing the excessive and rapid deterioration of magnesium alloys [37–40]. As shown in Figure 2, the number of scholarly works on magnesium alloy corrosion has increased dramatically during the previous two decades. However, the number of articles begins to decline in 2017, indicating that certain limitations in magnesium corrosion prevention may exist. In comparison, the number of publications on bioinspired surface design has increased significantly over the last 20 years, demonstrating that bioinspired surface design has garnered a lot of attention.

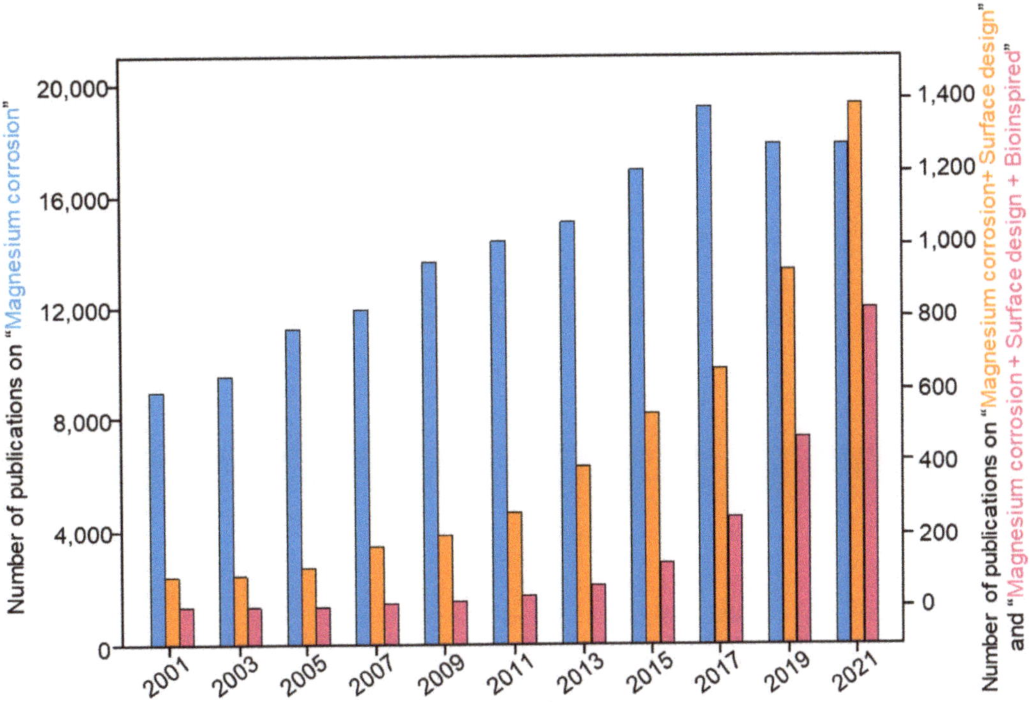

Figure 2. Publications related to magnesium alloy studies from 2001 to 2021. The data were obtained from the Web of Science database with the search date 20 June 2022, and the keywords were "Magnesium corrosion", "Magnesium corrosion + Surface design", and "Magnesium corrosion + Surface design + Bioinspired", respectively.

3. Bioinspired Solid Surface Design

Bioinspired solid-based surface treatments on magnesium alloys not only serve as anticorrosion design techniques, but also increase biocompatibility. For corrosion prevention, several solid surface-based protective barriers are used, including the bioinspired surface structure and bioinspired surface modification.

3.1. Bioinspired Surface Structure

Several research groups sought to build a biomimetic structure of a lotus leaf to generate a superhydrophobic surface on magnesium alloys, which is a viable approach for improving corrosion resistance [41–45] (Figure 3). The process is based on the perception that the surface air film can keep magnesium alloys from coming into direct contact with external aqueous solution [46,47]. Due to the tiny solid–liquid interfacial area on top of the protrusions, the liquid on top of this bioinspired surface is separated by trapped air. This can greatly limit the interaction between the corrosive species and the magnesium alloys and display the anticorrosion property. Several ways for endowing the bioinspired surface structure of magnesium alloys with corrosion resistance functions have been demonstrated [48–52].

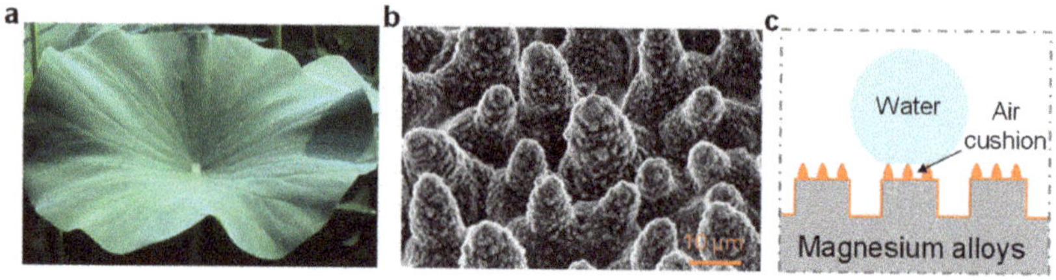

Figure 3. (**a**) Digital and (**b**) SEM images of a lotus leaf [53]; (**c**) illustration of contact modes of superhydrophobic surfaces under the Cassie state.

The bioinspired surface structure can be obtained by eliminating layers from the surface of magnesium alloys using the etching process [50,54,55]. The air regions on the bioinspired surface cannot be replaced by corrosive species, and the "air" parts of the surface are deemed fully non-wetting, which increases the corrosion resistance of magnesium alloys. Liu et al. [56] created a bioinspired surface by assembling peony-like micro-nano-size hierarchical structures on magnesium alloy surfaces and altering them in an ethanol solution containing fluorosilane for 12 h at ambient temperature. Low sliding angles (<5°) made it easy for water droplets to readily slide off. After 180 days of air exposure, the resulting surface's water contact angle changes very minimally, indicating a stable bioinspired surface structure and strong corrosion resistance. Liang et al. [57] created a bioinspired surface structure on AZ31 magnesium alloys with a contact angle (CA) of 154° and a sliding angle (SA) of 7.1° (Figure 4). When compared to the bare specimen, the bioinspired surface's corrosion potential E_{corr} is moved positively from −1.59 to −1.16 V, boosting its corrosion resistance. After 12 h of immersion in 3.5% NaCl solution, the charge transfer of the bioinspired surface is larger than that of the untreated magnesium alloys, exhibiting the long-term durability in NaCl solution.

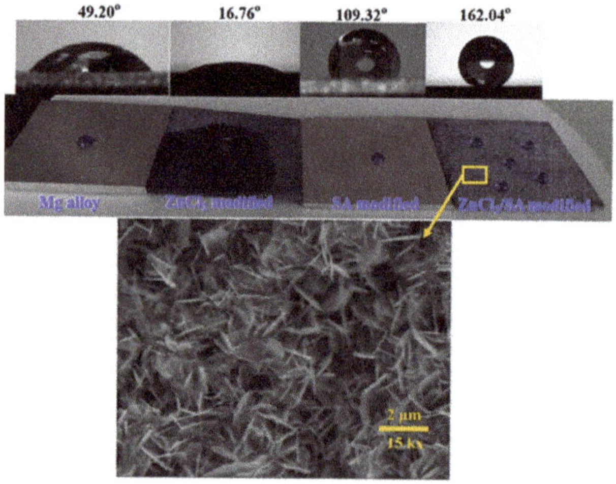

Figure 4. Water contact angles on the bioinspired surface of magnesium alloys [57].

The electrodeposition method of synthesis enables the creation of bioinspired surface structures with clearly defined morphologies and variable sizes, merely by adjusting the experimental synthesis parameters without the use of a template [58,59]. It has also been applied to the creation of bioinspired surfaces on magnesium alloys. Liu et al. [60] used an electrodeposition approach coupled with immersion in stearic acid solution to create a stearic acid/CeO_2 bioinspired surface on AZ31B magnesium alloys. With a high water contact angle (more than 158°) and low sliding angles (less than 2°), the modified surface demonstrated good superhydrophobic and self-cleaning properties. The I_{corr} for the bioinspired surface is 1.14×10^{-6} A cm^{-2} compared to 4.7×10^{-4} A cm^{-2} for the untreated surface, suggesting that the bioinspired surface greatly enhanced the magnesium alloys' anticorrosion property. The I_{corr} of the bioinspired surface increased to 1.56×10^{-4} A cm^{-2} even after 100 h of immersion in corrosive solution (Figure 5a), which is still lower than the untreated surface and demonstrates anticorrosion performance. Liu et al. [61] used one-step electrodeposition to create a bioinspired surface structure on Mg-Mn-Ce magnesium alloys. Compared to the value for the substrate, the I_{corr} of the bioinspired surface is more than two orders of magnitude lower. The bioinspired surface considerably improved the magnesium alloys' anticorrosion property in NaCl solution, as shown by the increase in E_{corr} and decrease in I_{corr} (Figure 5b). It was demonstrated that the CA of the bioinspired surface maintained above 150° after 400 mm abrasion by dragging the bioinspired surface on 1000# sandpaper in one direction at a pressure of 1.3 kPa, while CA of the bioinspired surface was lowered to 147.91° after abrasion for 500 mm, showing good mechanical durability to some extent.

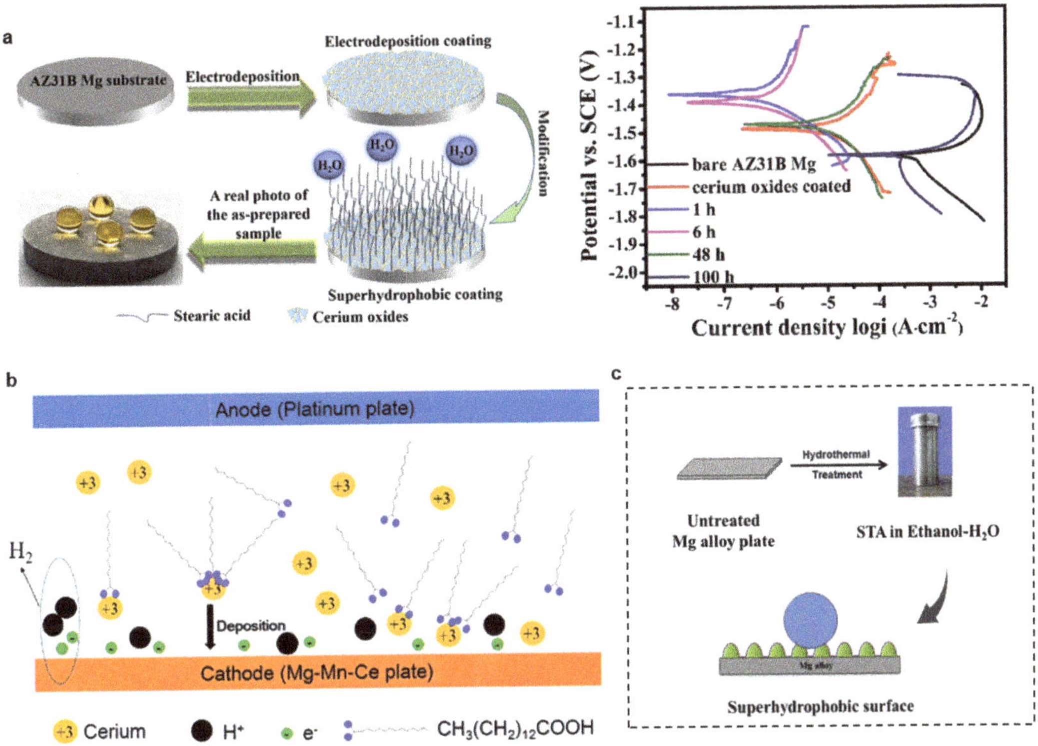

Figure 5. (**a**) The preparation process of the bioinspired surface structure prepared on the magnesium alloys' substrate and Tafel Polarization curves of the magnesium alloys immersed in NaCl solution for different times [60]; (**b**) Schematic illustration of the electrodeposition process [61]; and (**c**) The fabrication of the bioinspired surface structure on magnesium alloys using the hydrothermal process [62].

The hydrothermal synthesis approach typically employs a sealed system to develop a bioinspired surface structure on the magnesium alloys' surface by managing the controllable parameters. Feng et al. [62] utilized a one-step hydrothermal synthesis approach to create a bioinspired surface structure on AZ91 magnesium alloy (Figure 5c). After hydrothermal treatment, the I_{corr} drops from 1.48×10^{-4} A cm^{-2} to 3.32×10^{-7} A cm^{-2}. The

I_{corr} of the bioinspired surface drops by nearly three orders of magnitude when compared to the unmodified surface, demonstrating the bioinspired surface's better anticorrosion property in NaCl solution. Only a minor quantity of corrosion products was discovered on the bioinspired surface of the magnesium alloy after soaking in NaCl solution for 20 h, showing that the bioinspired surface was mildly damaged and could significantly increase corrosion resistance.

3.2. Bioinspired Surface Modification

Compared with the bioinspired surface structure, bioinspired surface modification has been widely used as a protective surface for magnesium alloys owing to its affordability, versatility, and applications [63]. A quick and low-cost approach for creating the anticorrosion coating on magnesium alloys was proposed by Li et al. [64]. With a CA of 154° and a SA of 7.1°, this bioinspired surface displayed exceptional superhydrophobicity. When compared to the unmodified magnesium alloys, the I_{corr} of the bioinspired surface significantly drops from 1.65×10^{-4} A cm^{-2} to 1.73×10^{-6} A cm^{-2}, or around two orders of magnitude (Figure 6). After being immersed in 3.5 weight percent NaCl for 336 h, the capacitive loop's diameter is noticeably larger than that of the unmodified magnesium alloys, demonstrating that the bioinspired surface modification dramatically increases the magnesium alloys' anticorrosion performance.

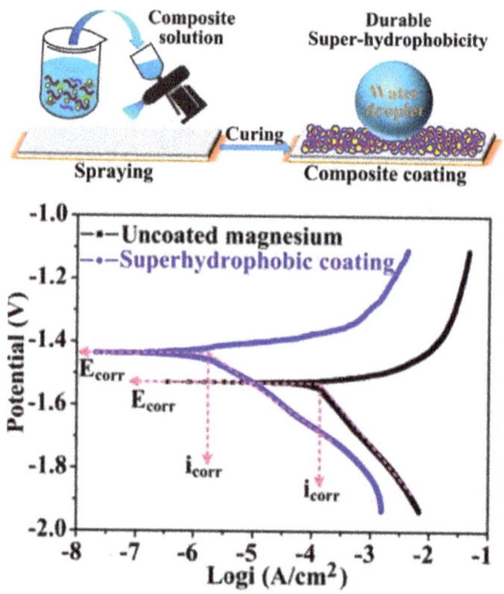

Figure 6. Surface modification process and corrosion resistance property of the bioinspired surface [64].

Although many bioinspired solid surfaces have been proposed for various biomedical uses, only a few have been used in clinical settings due to their poor biocompatibility. With inspiration from the natural ceramic structure, a high-quality ceramic surface is designed onto magnesium alloys, combining the mechanical strength of the metal with the superior biological properties of ceramics. This surface design can not only minimize the rate of corrosion of magnesium alloys in the biological environment, but it can also incorporate materials that can encourage bone formation to reduce cytotoxicity in the membrane layer and make it biologically active. This is a very promising bioinspired surface design for medical implantable metal materials.

The introduction of multi-functionalities through the use of bioactive coatings is another appealing technique to increase corrosion protection's effectiveness [65–68]. Bioactive coatings, including polylactic acid (PLA), polycaprolactone (PCL), chitosan (CS), polydopamine (PDA), and other biopolymers, are widely used in biomedical applications [69]. It can provide an adsorption center through electronic atoms and deposit magnesium alloy surfaces to hinder surface corrosion [70–74]. Construction of bioactive coatings can be realized by molecular interaction, such as electrostatic interaction or intermolecular force between functional molecular groups and the surface of the magnesium alloy. Marine mussels can cling to "non-adherent" magnesium alloys for corrosion protection because their foot protein has a high concentration of the amino acids, lysine and 3, 4-dihydroxy-L-phenylalanine [75,76]. By using layer-by-layer assembly on AZ31 magnesium alloy, Wang et al. [77] created a nanocomposite bioinspired surface. The formation of an abnormally dense layer adjacent to the surface, with a thickness of around 100 nm, results in the material having a high degree of protection against external molecules such as water or corrosive ions (Figure 7a).

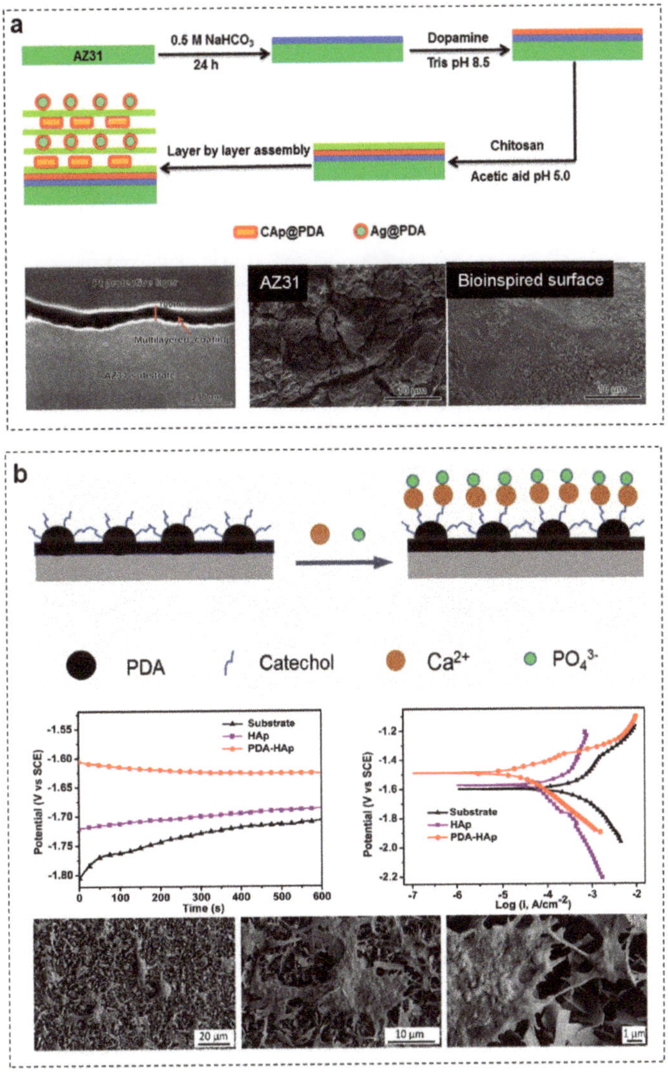

Figure 7. (**a**) Bioinspired surface modification of magnesium alloys using layer-by-layer assembly inspired by mussels, along with a cross-sectional image and the corroded surface morphology after polarization tests [77]; (**b**) Formation mechanism of bioinspired surface induced by polydopamine with corrosion protection abilities using electrochemical tests and cell adhesion on bioinspired surface [65].

To further improve cell adhesion, proliferation, and migration, the bioinspired polydopamine surface can also operate as a flexible substrate for the following surface-mediated reactions [78–81]. For instance, Zhou et al. [65] created a bioinspired protective surface on AZ31 magnesium alloy based on the hydrothermal treatment of hydroxyapatite and polydopamine. Compared to bare AZ31 (I_{corr} = 3.81 × 10^{-4} A cm^{-2}), the bioinspired surface showed the smallest I_{corr} (2.47 × 10^{-6} A cm^{-2}) and a high E_{corr} (−1.50 V), indicating a lower corrosion rate. Due to the immobilization of hydroxyapatite, the bioinspired surface promoted the proliferation of osteoblasts and the cell survival rate reached 120%. Therefore, the bioinspired surface is more suitable for the growth and proliferation of osteoblasts (Figure 7b).

In order to better improve the preparation, design, and performance, the layer-by-layer preparation approach extends the bioinspired solid surface design from 2D to 3D [82,83]. Chu et al. [84] proposed a comparable structure on Mg-Zn(6 wt%)-Ca(0.5 wt%) magnesium alloy modified with decreased graphene oxide/bistriethoxysilylethane composites for corrosion protection, drawing inspiration from the "bricks and mortar" structure of nacre. Due to the improved barrier effect and less galvanic corrosion caused by reduced porosity and flaws, the bioinspired surface has a high continuity and integrity and corrodes at a pace that is in orders of magnitude lower than that of naked magnesium alloy (Figure 8). The wear rate of untreated magnesium alloy is high, which is 3.5 × 10^{-3} mm^3 N^{-1} m^{-1}. It remarkably decreases to 3.98 × 10^{-5} mm^3 N^{-1} m^{-1} with the bioinspired surface, indicating the good anti-wear protection. Luo et al. [85] employed tannic acid containing Mg^{2+} ions to create a bioinspired surface on AZ31 magnesium alloy for better corrosion protection and high biocompatibility, which was inspired by the "tea stain" development process. Pan et al. [86] presented a chemically built bioinspired surface mixed with graphene oxide and heparin that worked as a compact barrier against corrosion spread. The multilayer coating can also greatly minimize hemolysis and platelet adhesion and activation, resulting in good blood compatibility.

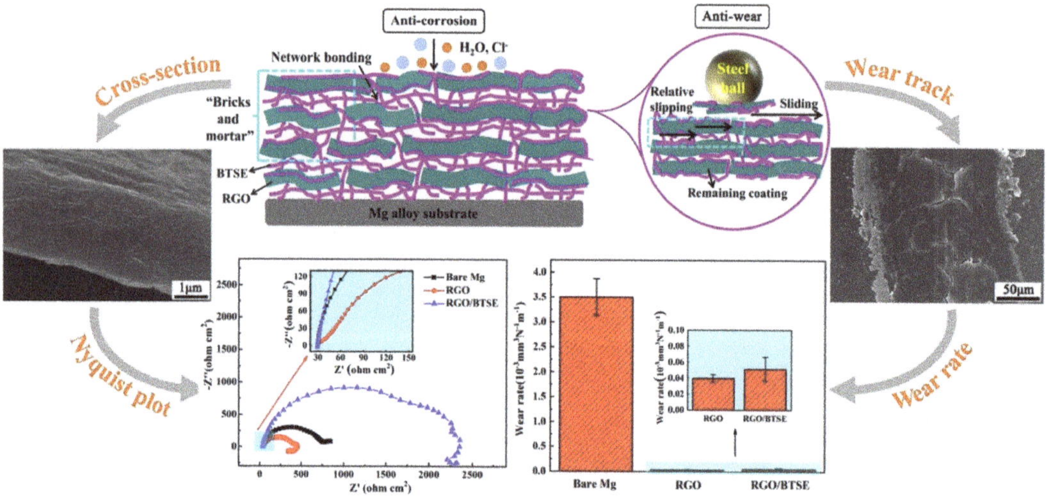

Figure 8. Mechanism of anticorrosion by bioinspired surface [84].

Bioinspired solid surfaces provide effective corrosion protection for magnesium alloys because a barrier layer of air film can keep magnesium alloys from coming into direct touch with external aqueous solution. It also can greatly limit the adsorption of bovine serum albumin and fully inhibit platelet adhesion and activation by influencing protein adsorption. However, only a handful have been used in clinical settings, since the single function of bioinspired solid surfaces is incompatible with biomedical applications.

For example, superhydrophobic surfaces, which are often utilized to increase magnesium alloys' anticorrosion property, are not favorable to cell attachment or cell growth. The optimal biomedical magnesium alloys should be multifunctional, enhancing not just anticorrosion characteristics, but also cell adhesion and proliferation. In fact, integrating many bioinspired capabilities onto a solid surface is rather difficult. That is, bioinspired solid surfaces still have a long way to go before they can be considered anticorrosive with good biocompatibility.

4. Bioinspired Liquid Surface Design

The specific properties of liquid interfaces operate as a physical barrier to increase long-term corrosion resistance [87]. The bioinspired liquid surface design of magnesium alloys is inspired by the natural Nepenthes pitcher plant, which can successfully isolate substrate corrosion by forming a stable immobilized liquid overlayer. Due to the capillary force and the van der Waals force, liquids are mobile and have unpredictable shapes on a solid surface, and they can be filled in micro-/nanoscale structures [88,89]. Three conditions must be met when creating magnesium alloys with a liquid surface [90]:

1. The functional liquid must be stabilized on the magnesium alloys' surfaces;
2. The functional liquid must preferentially wet the surfaces of the magnesium alloys over an aqueous solution;
3. The functional liquid and aqueous solution must be immiscible.

4.1. Bioinspired Surface Structure

To satisfy the first principle, the creation of micro-/nanostructures on magnesium alloys is employed to build the solid substrate with a tiny rough texture, improve the specific surface area, and enhance the storage of functional liquid [91,92]. Pant et al. [93] investigated the influence of several etching degrees of surface roughness on the stability of bioinspired liquid surfaces and discovered that a nanoscale roughness of 24.5 nm considerably improves functional liquid stability. Since surfaces with modest nanoscale roughness have higher adherence to functional liquids. Smaller and bigger rough surfaces exhibit functional liquid loss. The cloak effect easily occurs when there is insufficient functional liquid on the surface. Nanoscale roughness may be more favorable to the formation of stable functional liquid on magnesium alloy surfaces without im-pairing droplet sliding. On the one hand, nanoscale voids offer capillary force to lock the functional liquid without interfering with droplet mobility on the surface owing to excessive roughness. When compared to micrometer-scale holes, nanoscale rough structures generate more dense synaptic sites that can withstand the influence of external media and improve stability.

4.2. Bioinspired Surface Modification

To satisfy the second principle, the magnesium alloys' surface and functional liquid must match the physicochemical property to form a stable working system. Pretreatment with specific low surface energy groups promotes functional liquid distribution and storage in micro-/nano structures [94]. Common modification compounds, such as silane coupling agent, which connects magnesium alloys via silica-hydroxyl and organic matter, improve locking-liquid capability. Common low surface energy compounds, such as fluorosilane and fluorocarbon compounds (C_2F_8), can reduce the magnesium alloys' surface energy. Meanwhile, fluorinated groups are compatible with fluorinated functional liquid, which can react with it to produce precursors or mutually soluble tiny molecules. The functional liquid can be held for a long time, protecting magnesium alloys against corrosion over the long term.

In order to demonstrate whether the aqueous solution does not replace the functional liquid, the theoretical model is introduced in terms of the minimization of a system's free energy [90]. Specifically, configuration 1 refers to the state where the magnesium alloys' surface is infused with functional liquid, and the aqueous solution is floating on top of it (E_1). Configurations 2 and 3 refer to the states where the magnesium alloys' surface is completely wetted by functional liquid (E_2) and aqueous solution (E_3), respectively. To discover whether the surface of magnesium alloys has a greater affinity for the functional liquid than the aqueous solution, one should satisfy $\Delta E_A = E_3 - E_1 > 0$ and $\Delta E_B = E_3 - E_2 > 0$ (Figure 9). ΔE_A and ΔE_B can be further expressed as:

$$\Delta E_A = R(\gamma_f \cos\theta_f - \gamma_{aq} \cos\theta_{aq}) - \gamma_{faq}, \tag{7}$$

$$\Delta E_B = R(\gamma_f \cos\theta_f - \gamma_{aq} \cos\theta_{aq}) + \gamma_f - \gamma_{aq}, \tag{8}$$

where R is the magnesium alloys' surface roughness factor, γ_f the surface tension of the functional liquid, γ_{aq} the surface tension of aqueous solution, γ_{faq} the interfacial surface tension between the functional and aqueous solution, θ_f and θ_{aq} the contact angles of the functional liquid and aqueous solution on magnesium alloys' surface, and the subscripts f and aq the functional liquid and aqueous solution, respectively.

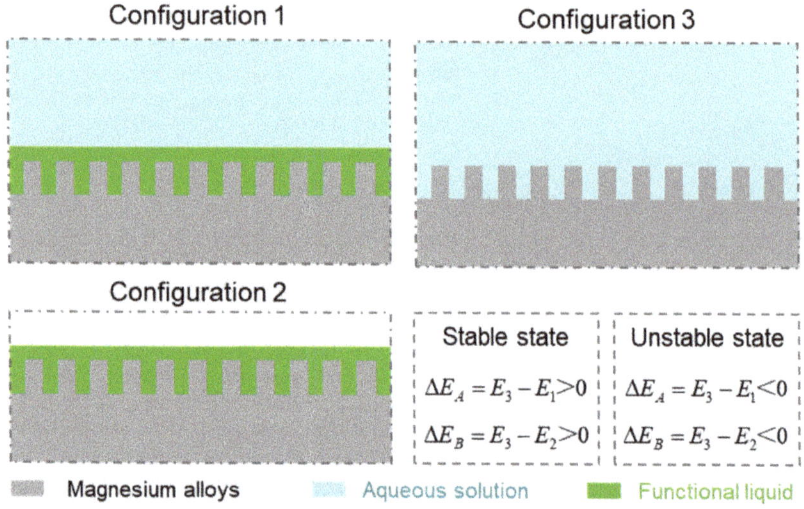

Figure 9. The working conditions for maintaining a stable liquid surface.

Preston et al. [95] and Anand et al. [96] analyzed and summarized four cases of unsuccessful designs in the interaction between the functional liquid and aqueous solution:

1. The surface tension of functional liquids is much lower than that of water, leading to water cloaking and the gradual loss of functional liquid by entraining as the droplet falls off the surface. The cloak effect is common when there is inadequate functional liquid on the surface. The spreading coefficient provides the cloaking criterion [95]:

$$S_{faq} = \gamma_{aq} - \gamma_{faq} - \gamma_f < 0, \tag{9}$$

where γ_{aq} denotes the surface tension of the aqueous solution; γ_{faq} the interfacial surface tension between the functional liquid and aqueous solution; γ_f the surface tension of the functional liquid, and the subscripts f and aq the functional liquid and aqueous solution, respectively.

2. The surface energy of the functional liquid is higher, so the aqueous solution can spread out on the functional liquid and cover the magnesium alloys' surface:

$$S_{aqf} = \gamma_f - \gamma_{aqf} - \gamma_{aq} < 0, \tag{10}$$

where γ_{aqf} denotes the interfacial surface tension between aqueous solution and functional liquid.

3. The functional liquid cannot completely infiltrate the magnesium alloys' surface.
4. The aqueous solution shows a higher affinity for the magnesium alloys' surface and can replace the functional liquid. Similarly, the spreading coefficient of functional liquid on magnesium alloys when in contact with air or aqueous solution can be calculated:

$$S_{fm} = \gamma_m - \gamma_{fm} - \gamma_f > -\gamma_f R, \tag{11}$$

$$S_{fm(aq)} = \gamma_{aqm} - \gamma_{fm} - \gamma_{faq} > -\gamma_{mf} R, \tag{12}$$

where γ_m denotes the surface energy of magnesium alloys, γ_{fm} the interfacial surface tension between the functional liquid and magnesium alloys, γ_{aqm} the interfacial surface tension between the aqueous solution and magnesium alloys, and the subscripts f, aq, and m the functional liquid, aqueous solution, and magnesium alloys, respectively. According to the three bioinspired liquid surface criteria, the magnesium alloys are closer to the functional liquid when $\gamma_{faq}(R-1)/(R-\varphi) < S_{fm(aq)} < 0$, where φ represents the fraction of the magnesium alloys' surface area filled by the functional liquid.

If criterion (9) is not met, the water droplet is "cloaked," or coated with a thin layer of functional liquid (Figure 10a), which can progressively deplete the functional liquid as water droplets depart. If condition (10) is not satisfied, the aqueous solution forms a film on top of the functional liquid (Figure 10b). To guarantee that the functional liquid remains infused in the magnesium alloys' surface, criteria (11) and (12) must be satisfied. If $S_{fm} \geq 0$ or $S_{fm(aq)} \geq 0$, the functional liquid thoroughly wets the magnesium alloys. Otherwise, if criteria (11) or criteria (12) are still met, but $S_{fm} < 0$ or $S_{fm(aq)} < 0$, a portion of the magnesium alloys comes into touch with the water droplets (Figure 10c,d). Therefore, it can be seen that the design of bioinspired liquid surfaces has harsh conditions. The rough structure that does not comply with the specifications or an inappropriate functional liquid cannot meet the constraints of the above formula. As a result, the stability or protective effect of the bioinspired liquid is reduced, resulting in the loss of functional liquid or erosion of the magnesium alloys by external aqueous solution.

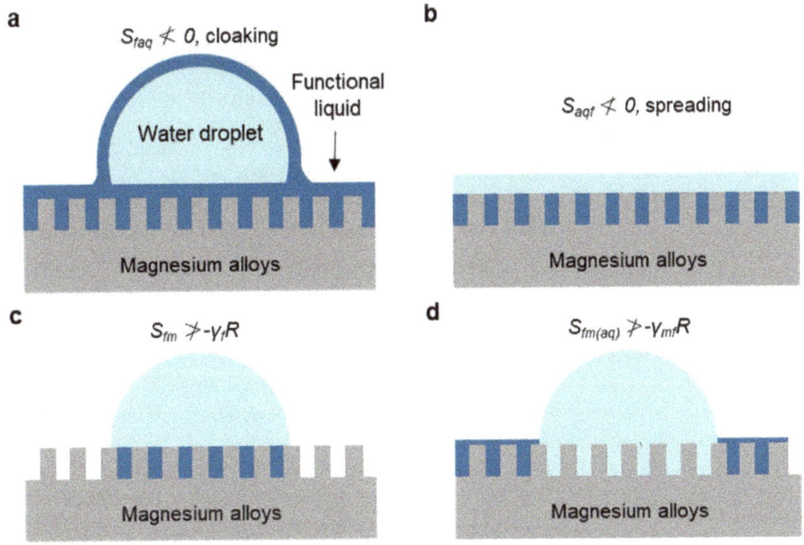

Figure 10. Water droplet put on a magnesium alloy's surface impregnated with a functional liquid that (**a**) thoroughly wets the magnesium alloys, (**b**) wets the magnesium alloys with a non-zero contact angle in the presence of air and the water droplet, and (**c**,**d**) functional liquid cannot remain infused in the magnesium alloy's surface.

4.3. Functional Liquid

To satisfy the third principle, the functional liquid used must be chosen carefully to guarantee that the final infused surface has appropriate qualities. With these fundamental characteristics in mind, it is evident that there is no optimal functional liquid for all scenarios. Rather, the selection of functional liquid depends on the application of the surface. The functional liquid often has the following characteristics [97]:

1. Low surface tension, allowing it to spread readily and penetrate roughness.

The low surface tension of the functional liquid is almost a necessary condition for bioinspired liquid surface design. The surface tension of the typical functional liquid is summarized in the literature [95,98]. Functional liquid with too low surface tension, on the other hand, is detrimental to interface stability. Sett et al. [98] observed that when perfluoropolyether oil and silicone oil come into contact with water droplets, a cloak occurs, resulting in functional liquid loss.

2. Low vapor pressure (<1 Pa), so as not to evaporate fast.

Functional liquid with low vapor pressure (perfluoropolyether krytox is 10^{-8} Pa, mineral oil 10^{-2} is Pa, silicone oil is 10^{-1} Pa) can be stored for a long time. However, the majority of functional liquids are composed mostly of carbon compounds, which can lead to the breakdown of functional liquid in high-temperature or UV-radiation environments. Zhang et al. [99] detected that the functional liquid progressively evaporates, and the bioinspired liquid surface loses its function at 65–75 °C, causing the surface contact angle to rise continually. In contrast, the ionic liquid is more stable than the perfluoropolyether and retains steady liquid surface performance under harsh environments such as 250 °C, strong UV irradiation, and high vacuum conditions.

3. Chemical inertness, in that it is not quickly destroyed when exposed to other chemicals.

Sett et al. [98] investigated the mutual solubility of flurane, silicone oil, mineral oil, and ionic liquid in different erosive liquids (such as water, ethanol, hexane, toluene, and ethylene glycol). According to studies, the chemical inertia of perfluoropolyether oil is the best, which does not react with external liquids, except for perfluorhexane with a comparable molecular structure. Ionic liquids, unlike non-polar fluorinated oil, silicone oil, and mineral oil, are not mutually soluble with water, but rather with alcohols, ethylene glycol, and toluene [100]. Except for water and perfluorohexane, silicone oil has the lowest chemical inertia and is nearly insoluble with other liquids. Even an inert functional liquid, however, can cause dissolution and failure due to molecular diffusion in the long-term immersion process of an ambient liquid [101]. Howell et al. [102] investigated and determined the solubility of krytox, a widely used perfluoropolyether functional liquid, to be 52 ± 46 ng cm^{-2} after 16 h of static immersion. As a result, the slow diffusion cannot be ignored for the seeming stability of the functional liquid layer.

4. Wide range of viscosity, with <100 cSt being the most typical; this value is neither too low to delay functional liquid depletion nor too high to hasten functional liquid infusion.

The viscosity of functional liquid influences the movement of the ambient medium on the functional liquid surface. When water droplets collide with the surface of the functional liquid, the interface shear force between the low viscosity functional liquid and the water droplets is negligible, allowing water droplet slide [103]. Too low viscosity, on the other hand, can compromise the stability of the functional liquid, leading to functional liquid loss.

To summarize, we believe that when selecting a functional liquid, elements such as solubility, surface energy, viscosity, and volatility should be considered extensively, and that a suitable functional liquid should be chosen in conjunction with the real service environment. Surface energy and viscosity that are too low do not promote the creation of a stable bioinspired liquid surface. In order to achieve a specified level of performance, a specific index of functional liquid must be used.

4.4. Application

The functional liquid on magnesium alloys possesses fluidity due to capillarity, which could fill in scratches and cure damage, as well as self-healing properties, of which the bioinspired liquid surface can keep the good super-slippery property under the condition of magnesium alloy damage, so as to avoid the loss of anti-corrosion and anti-biological adhesion [104–106]. Xiang et al. [107] prepared a superhydrophobic coating with a porous structure by means of electroplating and chemical replacement, and injected functional liquid on the surface to obtain a bioinspired liquid surface. It is found that the scratch narrowing of the bioinspired liquid surface damaged by the knife almost disappears after 2 h placement, which indicates that the functional liquid injected into the microporous surface can penetrate into the damaged cracks to plug the scratch, showing a good self-healing property. In addition, electrochemical testing and immersion in sodium chloride solution revealed that the bioinspired liquid surface of the scratch had self-healing properties.

Bioinspired liquid surface design also opens up new avenues for creating multi-functional surfaces that benefit from both long-term anticorrosion and biocompatibility. The essence of bioinspired liquid surface is that the air layer in micro/nano structures is replaced by functional liquid, which makes the water slide on the surface and cannot directly contact the magnesium alloys, giving the bioinspired liquid surface excellent corrosion resistance. Zhang et al. [44] created a two-layered surface for enhancing the anticorrosion performance of AZ31B magnesium alloy. The functional liquid was injected into the porous top layer, while the compact underlayer served as a corrosion barrier (Figure 11a). Magnesium alloys lower the I_{corr} of the useful liquid surface by six orders of magnitude. The solid superhydrophobic surface can operate for 3 days after soaking in sodium chloride solution, and the functional liquid surface can work for at least 15 days. The results reveal that the functional liquid surface outperforms the solid surface in terms of corrosion inhibition. Jiang et al. [108] created an anticorrosion system on AZ91D magnesium alloy that included a plasma electrolytic oxidation (PEO) film, hydroxide coating, and a functional liquid. Figure 11b depicts the anticorrosion system protection mechanism. The PEO coating formed in situ on the magnesium alloys' surface can provide a modest corrosion barrier. The sandwiched LDH film performed ternary functions, such as loading inhibitor, functional liquid anchoring, and PEO defect sealing. The functional liquid surface was water-repellent and self-reparable in the event of surface damage. Chloride ions can cause molybdate to be released in the LDH membrane, preventing further corrosion. This system can be immersed in sodium chloride solution for 20 days.

In the process of employing magnesium or magnesium alloys, constructing a bioinspired liquid surface can effectively solve the problem of corrosion resistance. Due to its fluidity, the bioinspired liquid surface can reduce the formation of defects and repair possible damage when compared to a bioinspired solid surface. In addition, it endows them with unique abilities, including antibiotic adhesion and antibacterial properties, to meet market applications that are always improving. The potential for bioinspired liquid surfaces in anticorrosion applications is anticipated to be the future study area in this field.

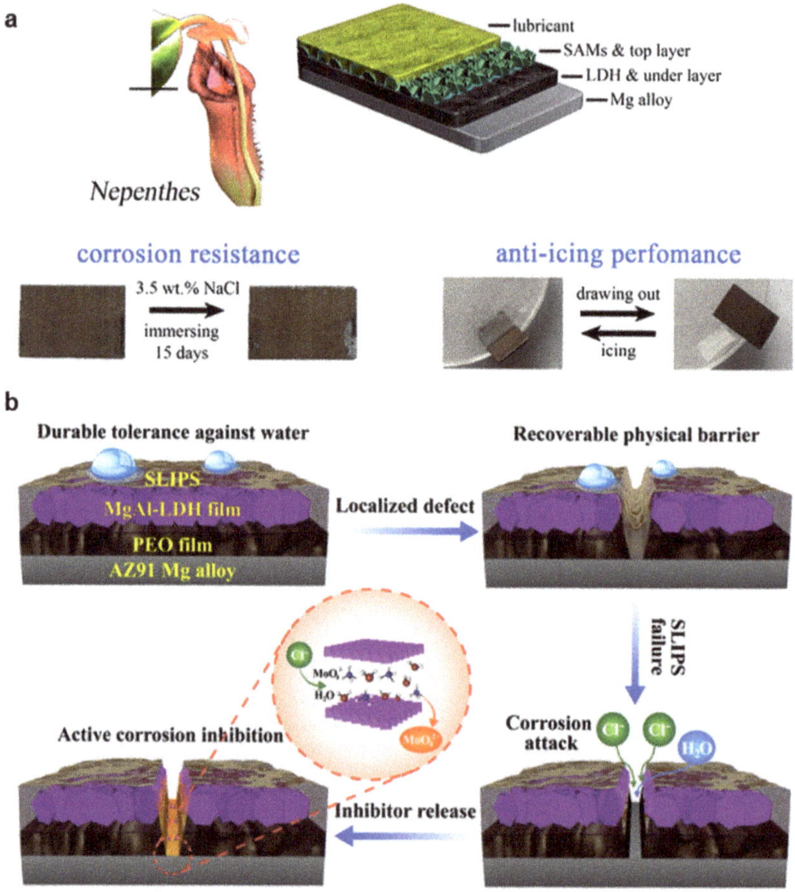

Figure 11. (**a**) A two-layered coating on magnesium alloys and its ability to resist corrosion [44]; (**b**) Schematic protection mechanism for anticorrosion on magnesium alloys [108].

5. Conclusions and Prospects

The use of bioinspired surface designs to increase magnesium alloys' anticorrosion performance is a successful strategy. Bioinspired solid surface design, including bioinspired surface structure and bioinspired surface modification, often generates a physical barrier on the surface of magnesium alloys to improve its anticorrosion performance. The use of bioinspired liquid surface design to produce a resilient and self-reparable functional liquid barrier paired with active corrosion inhibition endows the magnesium alloys with excellent anticorrosion performance. To date, bioinspired surface design can now increase not just the magnesium alloys' anticorrosion performance, but also their biocompatibility. However, it is very difficult to achieve a controllable corrosion rate of biodegradable, biomedical magnesium alloys as implant materials, since the protective qualities of various bioinspired surfaces on magnesium alloys are not well understood. As a result, systematic and long-term comparison and assessment are required in future study to explain the benefits and drawbacks of each bioinspired surface. Moreover, fabricating an ideal medical degradable magnesium alloy is still a challenge, in combination with corrosion resistance, self-degradation, biocompatibility, and drug release properties. Therefore, further study is needed to develop a novel bioactive surface for magnesium alloys. There is no doubt that bioinspired surface design advances both basic research and material therapy development.

Author Contributions: F.W. and C.P. planned the review, conducted the literature search, and authored the paper. Y.L. and J.X. assisted in accumulating extensive information on the issue and data presentation, as well as doing technical editing for all corrections. All authors have read and agreed to the published version of the manuscript.

Funding: This work was supported by the National Natural Science Foundation of China (31870952).

Institutional Review Board Statement: Not applicable.

Informed Consent Statement: Not applicable.

Data Availability Statement: In Figure 1, the data of publications related to magnesium alloy studies were obtained from the Web of Science database with the search date 20 June 2022.

Conflicts of Interest: The authors declare no conflict of interest.

References

1. Prasad, S.V.S.; Prasad, S.B.; Verma, K.; Mishra, R.K.; Kumar, V.; Singh, S. The role and significance of Magnesium in modern day research—A review. *J. Magnes. Alloy.* **2021**, *10*, 1–61. [CrossRef]
2. Park, J.E.; Jang, Y.S.; Choi, J.B.; Bae, T.S.; Park, I.S.; Lee, M.H. Evaluation of Corrosion Behavior and In Vitro of Strontium-Doped Calcium Phosphate Coating on Magnesium. *Materials* **2021**, *14*, 6625. [CrossRef] [PubMed]
3. Yang, Y.; Wu, Y.; Wei, Y.; Zeng, T.; Cao, B.; Liang, J. Preparation and Characterization of Hydroxyapatite Coating on AZ31 Magnesium Alloy Induced by Carboxymethyl Cellulose-Dopamine. *Materials* **2021**, *14*, 1849. [CrossRef] [PubMed]
4. Luo, Q.; Zhai, C.; Gu, Q.; Zhu, W.; Li, Q. Experimental study and thermodynamic evaluation of Mg–La–Zn system. *J. Alloys Compd.* **2020**, *814*, 152297. [CrossRef]
5. Staiger, M.P.; Pietak, A.M.; Huadmai, J.; Dias, G. Magnesium and its alloys as orthopedic biomaterials: A review. *Biomaterials* **2006**, *27*, 1728–1734. [CrossRef]
6. Hou, L.; Li, Z.; Pan, Y.; Du, L.; Li, X.; Zheng, Y.; Li, L. In vitro and in vivo studies on biodegradable magnesium alloy. *Prog. Nat. Sci. Mater. Int.* **2014**, *24*, 466–471. [CrossRef]
7. Li, N.; Zheng, Y. Novel magnesium alloys developed for biomedical application: A review. *J. Mater. Sci. Technol.* **2013**, *29*, 489–502. [CrossRef]
8. Jamel, M.; Lopez, H.; Schultz, B.; Otieno, W. The Effect of Solidification Rate on the Microstructure and Mechanical Properties of Pure Magnesium. *Metals* **2021**, *11*, 1264. [CrossRef]
9. Li, S.; Yang, X.; Hou, J.; Du, W. A review on thermal conductivity of magnesium and its alloys. *J. Magnes. Alloy.* **2020**, *8*, 78–90. [CrossRef]
10. Mordike, B.; Ebert, T. Magnesium: Properties-applications-potential. *Mater. Sci. Eng. A* **2001**, *302*, 37–45. [CrossRef]
11. Li, Z.; Yang, H.; Liu, J. Comparative study on yield behavior and non-associated yield criteria of AZ31B and ZK61 M magnesium alloys. *Mater. Sci. Eng. A* **2019**, *759*, 329–345. [CrossRef]
12. Saberi, A.; Bakhsheshi-Rad, H.R.; Abazari, S.; Ismail, A.F.; Sharif, S.; Ramakrishna, S.; Daroonparvar, M.; Berto, F. A comprehensive review on surface modifications of biodegradable magnesium-based implant alloy: Polymer coatings opportunities and challenges. *Coatings* **2021**, *11*, 747. [CrossRef]
13. Duygulu, O.; Kaya, R.A.; Oktay, G.; Kaya, A.A. In Investigation on the potential of magnesium alloy AZ31 as a bone implant. *Mater. Sci. Forum* **2007**, *546*, 421–424. [CrossRef]
14. Denkena, B.; Witte, F.; Podolsky, C.; Lucas, A. In Degradable implants made of magnesium alloys. In Proceedings of the 5th EUSPEN International Conference, Montpellier, France, 8–11 May 2005.
15. Rahim, M.I.; Ullah, S.; Mueller, P.P. Advances and challenges of biodegradable implant materials with a focus on magnesium-alloys and bacterial infections. *Metals* **2018**, *8*, 532. [CrossRef]
16. Liu, C.; Ren, Z.; Xu, Y.; Pang, S.; Zhao, X.; Zhao, Y. Biodegradable magnesium alloys developed as bone repair materials: A review. *Scanning* **2018**, *2018*, 9216314. [CrossRef]
17. Herber, V.; Okutan, B.; Antonoglou, G.; Sommer, G.N.; Payer, M. Bioresorbable magnesium-based alloys as novel biomaterials in oral bone regeneration: General review and clinical perspectives. *J. Clin. Med.* **2021**, *10*, 1842. [CrossRef]
18. Liu, C.; Xin, Y.; Tian, X.; Chu, P.K. Degradation susceptibility of surgical magnesium alloy in artificial biological fluid containing albumin. *J. Mater. Res.* **2007**, *22*, 1806–1814. [CrossRef]
19. Hsu, C.S.; Nazari, M.H.; Li, Q.; Shi, X. Enhancing degradation and corrosion resistance of AZ31 magnesium alloy through hydrophobic coating. *Mater. Chem. Phys.* **2019**, *225*, 426–432. [CrossRef]
20. Xin, Y.; Huo, K.; Tao, H.; Tang, G.; Chu, P.K. Influence of aggressive ions on the degradation behavior of biomedical magnesium alloy in physiological environment. *Acta Biomater.* **2008**, *4*, 2008–2015. [CrossRef]
21. Harandi, S.E.; Banerjee, P.C.; Easton, C.D.; Raman, R.S. Influence of bovine serum albumin in Hanks' solution on the corrosion and stress corrosion cracking of a magnesium alloy. *Mater. Sci. Eng. C* **2017**, *80*, 335–345. [CrossRef]
22. Gu, X.; Zheng, Y.; Chen, L. Influence of artificial biological fluid composition on the biocorrosion of potential orthopedic Mg–Ca, AZ31, AZ91 alloys. *Biomed. Mater.* **2009**, *4*, 065011. [CrossRef] [PubMed]

23. Höche, D.; Blawert, C.; Lamaka, S.V.; Scharnagl, N.; Mendis, C.; Zheludkevich, M.L. The effect of iron re-deposition on the corrosion of impurity-containing magnesium. *Phys. Chem. Chem. Phys.* **2016**, *18*, 1279–1291. [CrossRef] [PubMed]
24. Zhou, R.; Ming, Z.; He, J.; Ding, Y.; Jiang, J. Effect of magnesium hydroxide and aluminum hydroxide on the thermal stability, latent heat and flammability properties of Paraffin/HDPE phase change blends. *Polymers* **2020**, *12*, 180. [CrossRef]
25. Chen, X.B.; Zhou, X.; Abbott, T.B.; Easton, M.A.; Birbilis, N. Double-layered manganese phosphate conversion coating on magnesium alloy AZ91D: Insights into coating formation, growth and corrosion resistance. *Surf. Coat. Technol.* **2013**, *217*, 147–155. [CrossRef]
26. Noviana, D.; Paramitha, D.; Ulum, M.F.; Hermawan, H. The effect of hydrogen gas evolution of magnesium implant on the postimplantation mortality of rats. *J. Orthop. Transl.* **2016**, *5*, 9–15. [CrossRef]
27. Koch, K.; Bhushan, B.; Barthlott, W. Multifunctional surface structures of plants: An inspiration for biomimetics. *Prog. Mater. Sci.* **2009**, *54*, 137–178. [CrossRef]
28. Barthlott, W.; Neinhuis, C. Purity of the sacred lotus, or escape from contamination in biological surfaces. *Planta* **1997**, *202*, 1–8. [CrossRef]
29. Li, X.; Liu, S.; Du, Y. Investigation on the corrosion resistance of the Mg-10Al-xMn alloys based on thermodynamic calculations. *Corros. Sci.* **2021**, *189*, 109631. [CrossRef]
30. Song, G.L. Corrosion electrochemistry of magnesium (Mg) and its alloys. *Corros. Magnes. Alloy.* **2011**, *56*, 3–65.
31. Gerengi, H.; Cabrini, M.; Solomon, M.; Kaya, E. Understanding the Corrosion Behavior of the AZ91D Alloy in Simulated Body Fluid through the Use of Dynamic EIS. *ACS Omega* **2022**, *7*, 11929–11938. [CrossRef]
32. Baril, G.; Galicia, G.; Deslouis, C.; Pébère, N.; Tribollet, B.; Vivier, V. An impedance investigation of the mechanism of pure magnesium corrosion in sodium sulfate solutions. *J. Electrochem. Soc.* **2006**, *154*, C108. [CrossRef]
33. Song, G.; Atrens, A.; Stjohn, D.; Nairn, J.; Li, Y. The electrochemical corrosion of pure magnesium in 1 N NaCl. *Corros. Sci.* **1997**, *39*, 855–875. [CrossRef]
34. Abdalla, M.; Joplin, A.; Elahinia, M.; Ibrahim, H. Corrosion modeling of magnesium and its alloys for biomedical applications. *Corros. Mater. Degrad.* **2020**, *1*, 11. [CrossRef]
35. Narayanan, T.S.; Park, I.S.; Lee, M.H. Strategies to improve the corrosion resistance of microarc oxidation (MAO) coated magnesium alloys for degradable implants: Prospects and challenges. *Prog. Mater. Sci.* **2014**, *60*, 1–71. [CrossRef]
36. Zhang, W.; Jiang, Z.; Li, G.; Jiang, Q.; Lian, J. Electroless Ni-P/Ni-B duplex coatings for improving the hardness and the corrosion resistance of AZ91D magnesium alloy. *Appl. Surf. Sci.* **2008**, *254*, 4949–4955. [CrossRef]
37. Liu, Y.; Yin, X.; Zhang, J.; Yu, S.; Han, Z.; Ren, L. A electro-deposition process for fabrication of biomimetic super-hydrophobic surface and its corrosion resistance on magnesium alloy. *Electrochim. Acta* **2014**, *125*, 395–403. [CrossRef]
38. Cui, W.; Beniash, E.; Gawalt, E.; Xu, Z.; Sfeir, C. Biomimetic coating of magnesium alloy for enhanced corrosion resistance and calcium phosphate deposition. *Acta Biomater.* **2013**, *9*, 8650–8659. [CrossRef]
39. Fan, X.L.; Li, C.Y.; Wang, Y.B.; Huo, Y.F.; Li, S.Q.; Zeng, R.C. Corrosion resistance of an amino acid-bioinspired calcium phosphate coating on magnesium alloy AZ31. *J. Mater. Sci. Technol.* **2020**, *49*, 224–235. [CrossRef]
40. Wang, L.; Aversa, R.; Houa, Z.; Tian, J.; Liang, S.; Ge, S.; Chen, Y.; Perrotta, V.; Apicella, A.; Apicella, D. Bioresorption Control and Biological Response of Magnesium Alloy AZ31 Coated with Poly-β-Hydroxybutyrate. *Appl. Sci.* **2021**, *11*, 5627. [CrossRef]
41. Darmanin, T.; Guittard, F. Recent advances in the potential applications of bioinspired superhydrophobic materials. *J. Mater. Chem. A* **2014**, *2*, 16319–16359. [CrossRef]
42. Cao, Y.; Salvini, A.; Camaiti, M. Current status and future prospects of applying bioinspired superhydrophobic materials for conservation of stone artworks. *Coatings* **2020**, *10*, 353. [CrossRef]
43. Wang, X.; Song, W.; Li, Z.; Cong, Q. Fabrication of superhydrophobic AAO-Ag multilayer mimicking dragonfly wings. *Chin. Sci. Bull.* **2012**, *57*, 4635–4640. [CrossRef]
44. Zhang, J.; Gu, C.; Tu, J. Robust slippery coating with superior corrosion resistance and anti-icing performance for AZ31B Mg alloy protection. *ACS Appl. Mater. Interfaces* **2017**, *9*, 11247–11257. [CrossRef]
45. Zang, D.; Zhu, R.; Zhang, W.; Yu, X.; Lin, L.; Guo, X.; Liu, M.; Jiang, L. Corrosion-resistant superhydrophobic coatings on Mg alloy surfaces inspired by lotus seedpod. *Adv. Funct. Mater.* **2017**, *27*, 1605446. [CrossRef]
46. Ishizaki, T.; Masuda, Y.; Sakamoto, M. Corrosion resistance and durability of superhydrophobic surface formed on magnesium alloy coated with nanostructured cerium oxide film and fluoroalkylsilane molecules in corrosive NaCl aqueous solution. *Langmuir* **2011**, *27*, 4780–4788. [CrossRef]
47. Chu, J.; Tong, L.; Wen, M.; Jiang, Z.; Wang, K.; Zhang, H. Graphene oxide film as a protective barrier for Mg alloy: Worse or better is dependent on a chemical reduction process. *Carbon* **2019**, *145*, 389–400. [CrossRef]
48. Jiang, D.; Zhou, H.; Wan, S.; Cai, G.Y.; Dong, Z.H. Fabrication of superhydrophobic coating on magnesium alloy with improved corrosion resistance by combining micro-arc oxidation and cyclic assembly. *Surf. Coat. Technol.* **2018**, *339*, 155–166. [CrossRef]
49. Zhao, Y.; Shi, L.; Ji, X.; Li, J.; Han, Z.; Li, S.; Zeng, R.; Zhang, F.; Wang, Z. Corrosion resistance and antibacterial properties of polysiloxane modified layer-by-layer assembled self-healing coating on magnesium alloy. *J. Colloid Interface Sci.* **2018**, *526*, 43–50. [CrossRef]
50. Wu, Y.; Wang, Y.; Liu, H.; Liu, Y.; Guo, L.; Jia, D.; Ouyang, J.; Zhou, Y. The fabrication and hydrophobic property of micro-nano patterned surface on magnesium alloy using combined sparking sculpture and etching route. *Appl. Surf. Sci.* **2016**, *389*, 80–87. [CrossRef]

51. Yin, X.; Mu, P.; Wang, Q.; Li, J. Superhydrophobic ZIF-8-based dual-layer coating for enhanced corrosion protection of Mg alloy. *ACS Appl. Mater. Interfaces* **2020**, *12*, 35453–35463. [CrossRef]
52. Zhang, J.; Wei, J.; Li, B.; Zhao, X.; Zhang, J. Long-term corrosion protection for magnesium alloy by two-layer self-healing superamphiphobic coatings based on shape memory polymers and attapulgite. *J. Colloid Interface Sci.* **2021**, *594*, 836–847. [CrossRef] [PubMed]
53. Ensikat, H.J.; Ditsche-Kuru, P.; Neinhuis, C.; Barthlott, W. Superhydrophobicity in perfection: The outstanding properties of the lotus leaf. *Beilstein J. Nanotechnol.* **2011**, *2*, 152–161. [CrossRef] [PubMed]
54. Liu, Y.; Yin, X.; Zhang, J.; Wang, Y.; Han, Z.; Ren, L. Biomimetic hydrophobic surface fabricated by chemical etching method from hierarchically structured magnesium alloy substrate. *Appl. Surf. Sci.* **2013**, *280*, 845–849. [CrossRef]
55. Zhu, J.; Wan, H.; Hu, X. A rapid one-step process for the construction of corrosion-resistant bionic superhydrophobic surfaces. *Prog. Org. Coat.* **2016**, *100*, 56–62. [CrossRef]
56. Liu, K.; Zhang, M.; Zhai, J.; Wang, J.; Jiang, L. Bioinspired construction of Mg–Li alloys surfaces with stable superhydrophobicity and improved corrosion resistance. *Appl. Phys. Lett.* **2008**, *92*, 183103. [CrossRef]
57. Liang, M.; Wei, Y.; Hou, L.; Wang, H.; Li, Y.; Guo, C. Fabrication of a super-hydrophobic surface on a magnesium alloy by a simple method. *J. Alloys Compd.* **2016**, *656*, 311–317. [CrossRef]
58. Tonelli, D.; Scavetta, E.; Gualandi, I. Electrochemical deposition of nanomaterials for electrochemical sensing. *Sensors* **2019**, *19*, 1186. [CrossRef]
59. Chang, X.; Li, M.; Tang, S.; Shi, L.; Chen, X.; Niu, S.; Zhu, X.; Wang, D.; Sun, S. Superhydrophobic micro-nano structured PTFE/WO3 coating on low-temperature steel with outstanding anti-pollution, anti-icing, and anti-fouling performance. *Surf. Coat. Technol.* **2022**, *434*, 128214. [CrossRef]
60. Liu, X.; Zhang, T.C.; He, H.; Ouyang, L.; Yuan, S. A stearic Acid/CeO2 bilayer coating on AZ31B magnesium alloy with superhydrophobic and self-cleaning properties for corrosion inhibition. *J. Alloys Compd.* **2020**, *834*, 155210. [CrossRef]
61. Liu, Q.; Chen, D.; Kang, Z. One-step electrodeposition process to fabricate corrosion-resistant superhydrophobic surface on magnesium alloy. *ACS Appl. Mater. Interfaces* **2015**, *7*, 1859–1867. [CrossRef]
62. Feng, L.; Zhu, Y.; Wang, J.; Shi, X. One-step hydrothermal process to fabricate superhydrophobic surface on magnesium alloy with enhanced corrosion resistance and self-cleaning performance. *Appl. Surf. Sci.* **2017**, *422*, 566–573. [CrossRef]
63. Yu, D.; Qiu, H.; Mou, X.; Dou, Z.; Zhou, N.; Guo, Q.; Lyu, N.; Lu, L.; Yang, Z.; Huang, N. One-pot but two-step vapor-based amine- and fluorine-bearing dual-layer coating for improving anticorrosion and biocompatibility of magnesium alloy. *ACS Biomater. Sci. Eng.* **2019**, *5*, 4331–4340. [CrossRef] [PubMed]
64. Li, D.W.; Wang, H.Y.; Liu, Y.; Wei, D.S.; Zhao, Z.X. Large-scale fabrication of durable and robust super-hydrophobic spray coatings with excellent repairable and anti-corrosion performance. *Chem. Eng. J.* **2019**, *367*, 169–179. [CrossRef]
65. Zhou, Z.; Zheng, B.; Gu, Y.; Shen, C.; Wen, J.; Meng, Z.; Chen, S.; Ou, J.; Qin, A. New approach for improving anticorrosion and biocompatibility of magnesium alloys via polydopamine intermediate layer-induced hydroxyapatite coating. *Surf. Interfaces* **2020**, *19*, 100501. [CrossRef]
66. Wu, F.; Li, J.; Zhang, K.; He, Z.; Yang, P.; Zou, D.; Huang, N. Multifunctional coating based on hyaluronic acid and dopamine conjugate for potential application on surface modification of cardiovascular implanted devices. *ACS Appl. Mater. Interfaces* **2016**, *8*, 109–121. [CrossRef]
67. Li, J.; Wu, F.; Zhang, K.; He, Z.; Zou, D.; Luo, X.; Fan, Y.; Yang, P.; Zhao, A.; Huang, N. Controlling molecular weight of hyaluronic acid conjugated on amine-rich surface: Toward better multifunctional biomaterials for cardiovascular implants. *ACS Appl. Mater. Interfaces* **2017**, *9*, 30343–30358. [CrossRef]
68. Wu, F. Bioinspired Design for Medical Applications. In *Biomaterials and Materials for Medicine*; CRC Press: Boca Raton, FL, USA, 2021; pp. 319–328.
69. Sun, J.; Zhu, Y.; Meng, L.; Chen, P.; Shi, T.; Liu, X.; Zheng, Y. Electrophoretic deposition of colloidal particles on Mg with cytocompatibility, antibacterial performance, and corrosion resistance. *Acta Biomater.* **2016**, *45*, 387. [CrossRef]
70. Shivakumar, M.; Dharmaprakash, M.S.; Manjappa, S.; Nagashree, K.L. Corrosion inhibition performance of lignin extracted from black liquor on mild steel in 0.5 m H_2SO_4 acidic media. *Port. Electrochim. Acta* **2017**, *35*, 351. [CrossRef]
71. Shahmoradi, A.R.; Talebibahmanbigloo, N.; Javidparvar, A.A.; Bahlakehd, G.; Ramezanzadeh, B. Studying the adsorption/inhibition impact of the cellulose and lignin compounds extracted from agricultural waste on the mild steel corrosion in HCl solution. *J. Mol. Liq.* **2020**, *304*, 112751. [CrossRef]
72. Guedes, L.A.; Bacca, K.G.; Lopes, N.F.; Costa, E.M. Tannin of Acacia mearnsii as green corrosion inhibitor for AA7075-T6 alluminum alloy in acidic medium. *Mater. Corros.* **2019**, *70*, 1288. [CrossRef]
73. Abdulmajid, A.; Hamidon, T.S.; Rahim, A.A.; Hussin, M.H. Tamarind shell tannin extracts as green corrosion inhibitors of mild steel in hydrochloric acid medium. *Mater. Res. Express* **2019**, *6*, 106579. [CrossRef]
74. AhadiParsa, M.; Mohammadloo, H.E.; Mirabedini, S.M.; Roshanc, S. Bio-corrosion assessment and surface study of hydroxyapatite-coated AZ31 Mg alloy pre-treated with vinyl triethoxy silane. *Mater. Chem.* **2022**, *287*, 126147.
75. Qian, B.; Zheng, Z.; Michailids, M.; Fleck, N.; Bilton, M.; Song, Y.; Li, G.; Shchukin, D. Mussel-inspired self-healing coatings based on polydopamine-coated nanocontainers for corrosion protection. *ACS Appl. Mater. Interfaces* **2019**, *11*, 10283–10291. [CrossRef] [PubMed]

76. Lee, H.; Dellatore, S.M.; Miller, W.M.; Messersmith, P.B. Mussel-inspired surface chemistry for multifunctional coatings. *Science* **2007**, *318*, 426–430. [CrossRef]
77. Wang, B.; Zhao, L.; Zhu, W.; Fang, L.; Ren, F. Mussel-inspired nano-multilayered coating on magnesium alloys for enhanced corrosion resistance and antibacterial property. *Colloids Surf. B Biointerfaces* **2017**, *157*, 432–439. [CrossRef]
78. Chen, L.; Li, J.; Chang, J.; Jin, S.; Wu, D.; Yan, H.; Wang, X.; Guan, S. Mg-Zn-Y-Nd coated with citric acid and dopamine by layer-by-layer self-assembly to improve surface biocompatibility. *Sci. China Technol. Sci.* **2018**, *61*, 1228–1237. [CrossRef]
79. Bahremand, F.; Shahrabi, T.; Ramezanzadeh, B. Development of a nanostructured film based on samarium (III)/polydopamine on the steel surface with superior anti-corrosion and water-repellency properties. *J. Colloid Interface Sci.* **2021**, *582*, 342. [CrossRef]
80. Carangelo, A.; Acquesta, A.; Monetta, T. In-vitro corrosion of AZ31 magnesium alloys by using a polydopamine coating. *Bioact. Mater.* **2019**, *4*, 71. [CrossRef]
81. Singer, F.; Schlesak, M.; Mebert, C.; Höhn, S.; Virtanen, S. Corrosion properties of polydopamine coatings formed in one-step immersion process on magnesium. *ACS Appl. Mater. Interfaces* **2015**, *7*, 26758. [CrossRef]
82. Jia, Z.; Xiu, P.; Roohani-Esfahani, S.I.; Zreiqat, H.; Xiong, P.; Zhou, W.; Yan, J.; Cheng, Y.; Zheng, Y. Triple-bioinspired burying/crosslinking interfacial coassembly strategy for layer-by-layer construction of robust functional bioceramic self-coatings for osteointegration applications. *ACS Appl. Mater. Interfaces* **2019**, *11*, 4447–4469. [CrossRef]
83. Bouville, F.; Maire, E.; Meille, S.; Moortèle, B.V.; Stevenson, A.J.; Deville, S. Strong, tough and stiff bioinspired ceramics from brittle constituents. *Nat. Naterials* **2014**, *13*, 508. [CrossRef] [PubMed]
84. Chu, J.; Tong, L.; Wang, W.; Jiang, Z.; Sun, G.; Zou, D.; Wang, K.; Zhang, H. Sequentially bridged biomimetic graphene-based coating via covalent bonding with an effective anti-corrosion/wear protection for Mg alloy. *Colloids Surf. A Physicochem. Eng. Asp.* **2021**, *610*, 125707. [CrossRef]
85. Zhang, B.; Yao, R.; Li, L.; Li, M.; Yang, L.; Liang, Z.; Yu, H.; Zhang, H.; Luo, R.; Wang, Y. Bionic Tea Stain–Like, All-Nanoparticle Coating for Biocompatible Corrosion Protection. *Adv. Mater. Interfaces* **2019**, *6*, 1900899. [CrossRef]
86. Gao, F.; Hu, Y.; Li, G.; Liu, S.; Quan, L.; Yang, Z.; Wei, Y.; Pan, C. Layer-by-layer deposition of bioactive layers on magnesium alloy stent materials to improve corrosion resistance and biocompatibility. *Bioact. Mater.* **2020**, *5*, 611–623. [CrossRef] [PubMed]
87. Kan, Y.; Zheng, F.; Li, B.; Zhang, R.; Wei, Y.; Yu, Y.; Zhang, Y.; Ouyang, Y.; Qiu, R. Self-healing dual biomimetic liquid-infused slippery surface in a partition matrix: Fabrication and anti-corrosion capability for magnesium alloy. *Colloids Surf. A Physicochem. Eng. Asp.* **2021**, *630*, 127585. [CrossRef]
88. Li, H.; Feng, X.; Peng, Y.; Zeng, R. Durable lubricant-infused coating on a magnesium alloy substrate with anti-biofouling and anti-corrosion properties and excellent thermally assisted healing ability. *Nanoscale* **2020**, *12*, 7700–7711. [CrossRef]
89. Gao, S.; Li, X.; Zhang, M. Bionspired slippery surfaces by cluster-like ZnO@ Co3O4 and its anti-corrosion performance. *Dig. J. Nanomater. Biostruct. (DJNB)* **2021**, *16*, 1565–1573.
90. Wong, T.S.; Kang, S.H.; Tang, S.K.; Smythe, E.J.; Hatton, B.D.; Grinthal, A.; Aizenberg, J. Bioinspired self-repairing slippery surfaces with pressure-stable omniphobicity. *Nature* **2011**, *477*, 443–447. [CrossRef]
91. Yuan, S.; Zhang, X.; Lin, D.; Xu, F.; Li, Y.; Wang, H. A novel slippery surface with enhanced stability and corrosion resistance. *Prog. Org. Coat.* **2020**, *142*, 105563. [CrossRef]
92. Wu, D.; Ma, L.; Zhang, F.; Qian, H.; Minhas, B.; Yang, Y.; Han, X.; Zhang, D. Durable deicing lubricant-infused surface with photothermally switchable hydrophobic/slippery property. *Mater. Des.* **2020**, *185*, 108236. [CrossRef]
93. Pant, R.; Ujjain, S.K.; Nagarajan, A.K.; Khare, K. Enhanced slippery behavior and stability of lubricating fluid infused nanostructured surfaces. *Eur. Phys. J. Appl.* **2016**, *75*, 11301. [CrossRef]
94. Redon, R.; Vázquez-Olmos, A.; Mata-Zamora, M.; Ordóñez-Medrano, A.; Rivera-Torres, F.; Saniger, J. Contact angle studies on anodic porous alumina. *J. Colloid Interface Sci.* **2005**, *287*, 664–670. [CrossRef] [PubMed]
95. Preston, D.J.; Song, Y.; Lu, Z.; Antao, D.S.; Wang, E.N. Design of lubricant infused surfaces. *ACS Appl. Mater. Interfaces* **2017**, *9*, 42383–42392. [CrossRef] [PubMed]
96. Anand, S.; Paxson, A.T.; Dhiman, R.; Smith, J.D.; Varanasi, K.K. Enhanced condensation on lubricant-impregnated nanotextured surfaces. *ACS Nano* **2012**, *6*, 10122–10129. [CrossRef]
97. Peppou-Chapman, S.; Hong, J.K.; Waterhouse, A.; Neto, C. Life and death of liquid-infused surfaces: A review on the choice, analysis and fate of the infused liquid layer. *Chem. Soc. Rev.* **2020**, *49*, 3688–3715. [CrossRef]
98. Sett, S.; Yan, X.; Barac, G.; Bolton, L.W.; Miljkovic, N. Lubricant-infused surfaces for low-surface-tension fluids: Promise versus reality. *ACS Appl. Mater. Interfaces* **2017**, *9*, 36400–36408. [CrossRef]
99. Zhang, J.; Wu, L.; Li, B.; Li, L.; Seeger, S.; Wang, A. Evaporation-induced transition from nepenthes pitcher-inspired slippery surfaces to lotus leaf-inspired superoleophobic surfaces. *Langmuir* **2014**, *30*, 14292–14299. [CrossRef]
100. Pfruender, H.; Jones, R.; Weuster-Botz, D. Water immiscible ionic liquids as solvents for whole cell biocatalysis. *J. Biotechnol.* **2006**, *124*, 182–190. [CrossRef]
101. Vorobev, A. Dissolution dynamics of miscible liquid/liquid interfaces. *Curr. Opin. Colloid Interface Sci.* **2014**, *19*, 300–308. [CrossRef]
102. Howell, C.; Vu, T.L.; Johnson, C.P.; Hou, X.; Ahanotu, O.; Alvarenga, J.; Leslie, D.C.; Uzun, O.; Waterhouse, A.; Kim, P. Stability of surface-immobilized lubricant interfaces under flow. *Chem. Mater.* **2015**, *27*, 1792–1800. [CrossRef]
103. Kim, J.H.; Rothstein, J.P. Droplet impact dynamics on lubricant-infused superhydrophobic surfaces: The role of viscosity ratio. *Langmuir* **2016**, *32*, 10166–10176. [CrossRef] [PubMed]

104. Wang, X.; Gu, C.; Wang, L.; Zhang, J.; Tu, J. Ionic liquids-infused slippery surfaces for condensation and hot water repellency. *Chem. Eng. J.* **2018**, *343*, 561–571. [CrossRef]
105. Goodband, S.J.; Armstrong, S.; Kusumaatmaja, H.; Voïtchovsky, K. Effect of ageing on the structure and properties of model liquid-infused surfaces. *Langmuir* **2020**, *36*, 3461–3470. [CrossRef] [PubMed]
106. Zhang, M.; Chen, R.; Liu, Q.; Liu, J.; Yu, J.; Song, D.; Liu, P.; Gao, L.; Wang, J. Long-Term Stability of a Liquid-Infused Coating with Anti-Corrosion and Anti-Icing Potentials on Al Alloy. *ChemElectroChem* **2019**, *6*, 3911–3919. [CrossRef]
107. Xiang, T.; Zhang, M.; Sadig, H.R.; Li, Z.; Zhang, M.; Dong, C.; Yang, L.; Chan, W.; Li, C. Slippery liquid-infused porous surface for corrosion protection with self-healing property. *Chem. Eng. J.* **2018**, *345*, 147–155. [CrossRef]
108. Jiang, D.; Xia, X.; Hou, J.; Cai, G.; Zhang, X.; Dong, Z. A novel coating system with self-reparable slippery surface and active corrosion inhibition for reliable protection of Mg alloy. *Chem. Eng. J.* **2019**, *373*, 285–297. [CrossRef]

metals

Article

Comparative Investigation of the Corrosion Behavior and Biocompatibility of the Different Chemical Conversion Coatings on the Magnesium Alloy Surfaces

Lingjie Meng [1,†], Xuhui Liu [2,†], Li Liu [3,†], Qingxiang Hong [1], Yuxin Cheng [1], Fei Gao [4,*], Jie Chen [1], Qiuyang Zhang [1,*] and Changjiang Pan [1,*]

1 Faculty of Mechanical and Material Engineering, Jiangsu Provincial Engineering Research Center for Biomaterials and Advanced Medical Devices, Huaiyin Institute of Technology, Huai'an 223003, China
2 The Affiliated Huai'an Hospital of Xuzhou Medical University, Huai'an 223003, China
3 Department of Cardiology, The First College of Clinical Medical Sciences, China Three Gorges University, Yichang 443002, China
4 Chengdu Neurotrans Medical Technology Co., Ltd., Chengdu 610219, China
* Correspondence: phil.gao@neurotrans.com.cn (F.G.); qyzhang@hyit.edu.cn (Q.Z.); panchangjiang@hyit.edu.cn (C.P.)
† These authors contributed equally to this work.

Abstract: Due to their good biodegradability and biocompatibility, magnesium alloys are widely favored as the potential candidate for the biodegradable cardiovascular stent. However, the rapid degradation and the limited biocompatibility in vivo remain the main bottlenecks that inhibit their clinical applications. The construction of the chemical conversion coating on the magnesium alloy surface represents one of the effective strategies to control the degradation rate and enhance the biocompatibility. In the present study, the different chemical conversion layers were prepared on the magnesium alloy surface by chemical conversion treatment, including sodium hydroxide (NaOH), hydrofluoric acid (HF), phosphoric acid (H_3PO_4) and phytic acid ($C_6H_{18}O_{24}P_6$) treatment, and the corrosion behaviors and biocompatibility of the chemical conversion layers were comparatively investigated in detail. The results showed that the different chemical treatments can produce the different conversion layers on the magnesium alloy surfaces with a variety of physicochemical characteristics, corrosion resistance and biocompatibility, and all treatments can enhance the corrosion resistance to varying degrees. The hydrophilicity and corrosion resistance of the sodium hydroxide-treated magnesium alloy were the best among all the materials. Although the hydrofluoric acid-treated magnesium alloy had produced a hydrophobic coating, the corrosion resistance still needed to be improved. Magnesium alloys treated by sodium hydroxide showed a selective promotion of albumin adsorption, while the other samples simultaneously promoted albumin and fibrinogen adsorption. For the blood compatibility, the hemolysis rates of all of the treated materials were reduced to below 5%. The samples treated by phytic acid had the smallest hemolysis rate, and the NaOH-treated magnesium alloy had the least amount of platelet adhesion and activation. An appropriate microenvironment for cell growth could be achieved by the chemical conversion treatment, according to the results of the endothelial cell adhesion and proliferation, and the NaOH-treated surface showed the best endothelial cell growth behaviors among all of the samples. In summary, the corrosion resistance and biocompatibility of the magnesium alloy were significantly improved by the sodium hydroxide treatment, and thus this treatment can be used as a pretreatment for the surface modification of the magnesium alloy in order to further enhance the biocompatibility when used as the cardiovascular implants.

Keywords: magnesium alloy; surface modification; corrosion resistance; blood compatibility; cytocompatibility

check for updates

Citation: Meng, L.; Liu, X.; Liu, L.; Hong, Q.; Cheng, Y.; Gao, F.; Chen, J.; Zhang, Q.; Pan, C. Comparative Investigation of the Corrosion Behavior and Biocompatibility of the Different Chemical Conversion Coatings on the Magnesium Alloy Surfaces. *Metals* **2022**, *12*, 1644. https://doi.org/10.3390/met12101644

Academic Editor: Yadir Torres Hernández

Received: 19 August 2022
Accepted: 27 September 2022
Published: 30 September 2022

1. Introduction

Percutaneous coronary intervention (PCI) with the bare metal stents (BMS) or drug-eluting stents (DES) is the most popular way to treat stenotic coronary artery disease [1]. The insertion of the bare metal stents in the blood vessel can support and maintain the vessel lumen to maintain blood circulation [2]. The lower incidence of the intimal hyperplasia and in-stent restenosis is an advantage of DES over BMS [3]. Although conventional vascular stents made of non-degradable metals have the positive clinical outcomes and widespread applications, the stent will permanently remain in the body as foreign materials because they are non-degradable, which may cause a series of clinical side effects, such as the physical irritation over an extended period and the delayed endothelialization caused by chronic inflammation and neointimal atherosclerosis, etc. [4,5]. Therefore, the development of biodegradable stents in order to overcome the complications that exist with the permanent implants represents an important direction for the cardiovascular stents.

Magnesium and its alloys have attracted dominant interests as a novel and potential biodegradable metal for the cardiovascular stents because of its outstanding biocompatibility, biodegradability, and superior mechanical properties [6–10]. On the one hand, Mg^{2+} level in human blood ranges from 0.70 to 1.11 mmol/L and it is the fourth prevalent cation in the human body. Mg^{2+} could be absorbed by the body after the magnesium alloy implant degradation [4]. On the other hand, the chemical activity of the magnesium permits the in vivo degradation in the physiological environment, which can avoid secondary surgical removal and minimize the harm to the human body. Moreover, Mg also participates in protein synthesis and is essential to treat myocardial infarction and prevent neointimal atherosclerosis [11,12]. However, the high corrosion rate of magnesium and its alloy in vivo can cause a lot of adverse reactions, including excessive Mg^{2+} release [13], local alkalinization, hydrogen production [14] and the enrichment of the secondary corrosion products [15], which may cause the premature loss of the mechanical properties of the implant before completing its physiological function, finally leading to the implantation failure.

It is well known that both the corrosion resistance and the biocompatibility of magnesium alloys are closely related to surface characteristics [16]. The major challenges for the magnesium alloys in cardiovascular implants are to control the in vivo corrosion rate [17,18] and to enhance their surface endothelial cell growth for endothelialization [19]. These issues must be resolved for magnesium alloys to be used safely in clinical applications. One of the most effective ways to control the degradation rate and regulate the bioactivity of the magnesium alloys is surface modification. Currently, the main strategies for improving the corrosion resistance of the magnesium alloy can be roughly divided into three aspects: mechanical modification (friction wear [20] and shot peening [21], etc.), physical modification (ion implantation [22], magnetron sputtering [23], iron plating [24] and laser melting [25], etc.), and chemical modification (surface chemical transformation [26], anodic oxidation [27], micro-arc oxidation (MAO) [28] and surface self-assembly [29], etc.). A protective layer of metal oxides or other compounds can be formed on the surface using a variety of surface modification techniques, but chemical conversion treatment represents the most effective way to produce the protective layer which can physically isolate the magnesium substrate from the corrosive medium. Chemical conversion coating is frequently employed in the field of the biomedical devices because of its easy-operation and low cost-effectiveness [30,31]. It can not only improve the adhesion of the final deposited coating to the substrate, but also enhance the corrosion resistance and the biocompatibility [32]. One study coated magnesium alloys with a variety of chemical conversion layers, such as $Mg(OH)_2$ [33], fluoride [34], molybdate [35], phosphate [36] and rare earth salts [37]. Gao et al. [38] reported that $Mg(OH)_2$ films had the properties of high hardness, superior corrosion resistance, and good biocompatibility, but the thin and loose $Mg(OH)_2$ conversion layer could not prevent the penetration of corrosive fluids. In vitro and in vivo testing by Mao et al. [34] showed that the fluoride treatment considerably enhanced the corrosion resistance and biocompatibility of Mg alloy stent materials. Fluoride conversion films, however, were less bioactive and less stable after the prolonged immersion. The toxicity of molybdate conversion films has been reported to be unsuitable as the surface treatment for cardiovascular stent materials, despite their good corrosion resistance [35]. Although the phosphate coatings were insoluble in water and have outstanding biocompatibility, they were also very brittle and susceptible to breakdown [39]. The chemical stability and long-term insolubility of rare earth elements in physiological fluids were well established. The conversion film created by treating AZ31 magnesium alloy in cerium chloride solution was studied by Motemour et al. [40], and the results showed that it produced a thicker film with more coverage and superior corrosion resistance. Although a lot of the literature has reported that the chemical conversion layer can enhance the corrosion resistance and biocompatibility of the magnesium alloy, there is lack of the systematic comparative investigation between the different chemical treatments, especially when it is used for the cardiovascular implants.

AZ31B is a kind of widely used magnesium alloy for developing the cardiovascular implants. When the content of Al element in AZ31B magnesium alloy was about 3%, it could form sosoloid with magnesium, meanwhile Al could increase the strength and plasticity of magnesium alloy and improve the structure stability of the oxide film. Although Al is harmful to human body, the slow release of Al^{3+} after the implantation will not cause great harm to the human body. The goal of this paper is to compare the physicochemical properties, corrosion resistance, and biocompatibility of the surface chemical conversion films created by four common chemical treatments, including sodium hydroxide (NaOH), hydrofluoric acid (HF), phosphoric acid (H_3PO_4) and phytate ($C_6H_{18}O_{24}P_6$), as well as to screen the coating preparation process for the best corrosion resistance and biocompatibility to meet the clinical biological application. To this end, the above-mentioned four chemical treatments were applied to produce the chemical conversion layers on the magnesium alloy, and the physiochemical properties, corrosion resistance and biocompatibility were comparatively investigated in detail.

2. Materials and Methods

2.1. Sample Preparation

An AZ31B magnesium alloy rod with the diameter of 12 mm was firstly cut into 5 mm slices, and then polished to a mirror finish with 400#, 600#, 1200#, 1500#, and 2000# sandpapers in turn. The samples were ultrasonically cleaned in acetone and ethanol for 10 min, respectively, and marked as Mg.

Alkali heat treatment: the cleaned Mg samples were immersed into a 3 M NaOH solution to treat 24 h in a water bath at 75 °C. The samples were cleaned by the deionized water and then dried and labelled as Mg-OH.

Fluoridation treatment: in order to clean and remove the surface impurities, Mg was first immersed into a solution of 50 g/L NaOH and 10 g/L $Na_3PO_4 \cdot 12H_2O$ for 15 min at room temperature. The samples were cleaned and then immersed in a 12 mol/L HF solution to treat 15 min. After being cleaned and dried, the samples were recorded as Mg-HF.

Phosphating treatment: the polished magnesium alloy was placed into a 3 M NaOH solution for 30 min to remove the surface impurities such as oil and organic substances, and then the sample was cleaned by the deionized water and dried. Finally, the sample was put into a mixed solution of 0.2 g/L H_3PO_4, 0.3 g/L NaF_2 and 0.5 g/L BaH_2PO_4 to treat 20 min at 90 °C. The sample was cleaned and dried, and the as-prepared sample was recorded as Mg-P.

Phytic acid treatment: the magnesium alloy was first washed by a 3 M NaOH solution for 30 min to remove the surface contaminants followed by cleaned with the deionized water. After dried, the sample was placed into 0.7 wt% phytic acid ($C_6H_{18}O_{24}P_6$) solution (PH 5) for treating 40 min in a 60 °C water bath. The sample was cleaned and dried. The sample was recorded as Mg-PA.

2.2. Surface Characterization

The surface atomic concentrations and elemental binding states of the modified magnesium alloys were analyzed by X-ray photoelectron spectroscopy (XPS, Quantum 2000; PHI Co., Chanhassen, MN). The surface microscopic morphology of the sample was observed by a scanning electron microscopy (SEM, FEI Quanta250, United States). The water contact angle was measured by a water contact angle meter (Krüss GmbH, Germany) to characterize the surface hydrophilicity/hydrophobicity; three parallel samples were measured and the average value was calculated.

2.3. Electrochemical Corrosion Behaviors

2.3.1. Potentiodynamic Polarization Curve

The potentiodynamic polarization curves of the samples were measured by a standard three-electrode system using a CHI660D electrochemical workstation (CHI Instruments, Inc., Shanghai, China). The test solution was Hank's simulated body fluid (SBF, compositions: NaCl, 8 g/L; KCl, 0.4 g/L; $NaHCO_3$, 0.35 g/L; $CaCl_2$, 0.14 g/L; Na_2HPO_4, 0.06 g/L; KH_2PO_4, 0.06 g/L; $MgSO_4 \cdot 7H_2O$, 0.01 g/L; glucose, 1 g/L). The three-electrode system was composed of Ag/AgCl as the reference electrode, platinum wire as the auxiliary electrode, and the sample with an exposed area of 1 cm^2 as the working electrode. The sample was exposed to SBF solution for 10 min before the test to obtain a stable open circuit potential (OCP). The potentiodynamic polarization curve was measured at a scan rate of 1 mV/s and the corrosion potential (E_{corr}) and corrosion current density (i_{corr}) were obtained using the Tafel extrapolation method. The corrosion current density was used to calculate the annual corrosion depth according to the following formula [41]:

$$d = 3.28 \times 10^{-3}(M/n\rho)I_{corr} \tag{1}$$

where d is the corrosion depth (mm/y), M is the gram atomic weight of Mg (24 g/mol), n is the atomic valence of Mg (n = 2), ρ is the density of Mg (1.74 g/cm^3), and I_{corr} is the corrosion current density (μA/cm^2).

2.3.2. Immersion Experiment

To further evaluate the in vitro corrosion resistance of the blank and modified magnesium alloys, the samples were firstly sealed with the silicone rubber, exposing only 1 cm^2 area, and then the samples were immersed in 20 mL SBF solution and incubated at 37 ± 0.5 °C for 1, 3, 7 and 14 days, respectively. The SBF solution was changed every 2 days. After immersion, the samples were gently washed with the deionized water and dried in air. The surface corrosion morphologies and elemental compositions of the samples were observed by scanning electron microscopy (SEM, FEI Quanta250) and energy dispersive spectrometer (EDS) after the deposition of a gold layer, respectively.

2.3.3. pH Changes in Immersion Solution

The sealed samples with silicone rubber were immersed in 20 mL of pH 7.4 SBF solution for 1, 2, 3, 4, 5, 6 and 7 days, respectively, and the solution was changed every 2 days. The pH of the immersion solution was measured three times with a pH meter for each sample at the predetermined times and the average value was calculated. Finally, the pH change curve was plotted according to the values.

2.4. Protein Adsorption

The samples were firstly equilibrated with PBS solution for 2 h, then 2 mL of albumin and fibrinogen solution (1 mg/mL, 0.01 M PBS) were added, respectively. After fully adsorbed at 37 °C for 2.5 h, the samples were rinsed 3 times by PBS followed by adding 2 mL of sodium dodecyl sulfate solution (SDS, 1wt%) to ultrasonically desorb protein for 30 min. Finally, 100 μL of the solution was transferred into a 96-well plate, and the absorbance at 562 nm was measured. The protein adsorption amount was calculated according to the standard curve, three parallel samples were measured, and the values were averaged.

2.5. Blood Compatibility

2.5.1. Hemolysis Rate

The hemolysis rate test was performed according to the standard protocol of ISO 10993-4:2002. The healthy human whole blood containing 3.8 wt% sodium citrate was centrifuged 10 min at 1500× *g* rpm to obtain the red blood cells. The erythrocytes were diluted into 2% erythrocyte suspension with saline. The cleaned samples were placed into a 24-well plate, and then 2 mL of 2% erythrocyte suspension was added to each sample and incubated 3 h at 37 °C. Subsequently, 1 mL solution was centrifuged 5 min at 3000 rpm, and 200 μL of the supernatant was transferred into a new 96-well plate. The absorbance at 450 nm was measured by a microplate reader (Bio-Tek Eons). The 2% erythrocyte suspensions prepared with physiological saline and distilled water, respectively, were used as the negative and positive controls. The hemolysis rate was calculated as follows.

$$\text{Hemolysis (\%)} = (A - A_2)/(A_1 - A_2) \times 100\% \quad (2)$$

where A is the absorbance of the sample, A_1 is the absorbance of the positive control, A_2 is the absorbance of the negative control.

2.5.2. Platelet Adhesion

Platelet-rich plasma (PRP) was first obtained by centrifugating the fresh whole blood from a healthy volunteer at 1500× *g* rpm for 10 min. PRP was covered on the sample surface and incubated 2.5 h at 37 °C. Then, the non-adhered platelets were washed off by phosphate buffer solution (PBS) and the adherent platelets were fixed 3 h with 2.5% glutaraldehyde solution at 4 °C. Finally, the samples were immersed in 50%, 70%, 90%, and 100% ethanol solutions in turn for 15 min each and dried at room temperature. The adhered platelets were observed by scanning electron microscopy (SEM, FEI, Quanta250, United States) after spraying a gold layer. At the same time, five randomly selected SEM images (×3000) for each sample were used for statistical counting of the platelets.

2.6. Endothelial Cell Behaviors

2.6.1. Cell Adhesion

The samples sealed with silicone rubber were washed and placed into a 24-well cell culture plate for sterilizing 24 h under ultraviolet lamp. 0.5 mL of cell suspension (5×10^4 cells/mL, ECV304, Cobioer, Nanjing, China) and 1.5 mL of cell culture medium were added and incubated at 37 °C with 5% CO_2 in an incubator for 1 and 2 d, respectively. The cells were first washed twice with saline, and then fixed 3 h with 2.5% glutaraldehyde at 4 °C. After washed again, the adhered cells were stained successively by 100 μL of rhodamine (10 μg/mL) for 20 min and 100 μL of 4,6-diamidino-2-phenylindole (DAPI, 500 ng/mL) for 5 min. After washed and dried, the fluorescent pictures of the adhered cells were taken by an inverted fluorescence microscopy (Carl Zeiss A2 inverted) in the dark.

2.6.2. Cell Proliferation

The sealed samples were first placed in a 24-well plate and sterilized under UV light overnight. Then, 0.5 mL of 5×10^4 cells/mL endothelial cells and 1.5 mL of culture medium were added to each well and incubated for 1 and 2 d, respectively. Subsequently, the samples were transferred into a new 24-well plate, 0.5 mL of 10% CCK-8 solution was added to each sample and incubated 3.5 h at 37 °C. Finally, 200 μL of the solution was transferred into a 96-well plate, and the absorbance at 450 nm was measured with a microplate reader (Bio-Tek Eons). Three parallel samples were measured, and the values were averaged.

2.7. Statistical Analysis

All the data were expressed as mean ± standard deviation (SD) and statistically analyzed using SPSS 12.0. Statistically significant differences were determined by one-way analysis of variance (ANOVA), and $p < 0.05$ was considered to be statistically significant.

3. Results and Discussion

3.1. Surface Characterization

It is well known that the chemical treatment can produce the chemical conversion layers on the magnesium alloy surface concurrently with the introduction of new elements. In this study, the surface elemental concentrations and their valence states after the chemical treatments were firstly determined by X-ray photoelectron spectroscopy (XPS). Figure 1 and Table 1 show the survey spectra and elemental concentrations of the different samples, respectively. The high-resolution spectra of the different conversion layers are shown in Figure 2. The blank magnesium alloy was mainly made of the elements O, C, and Mg, indicating that a naturally occurring oxide layer can be produced due to the chemical activity of magnesium. The Mg_{1s} spectra had a distinctive peak of Mg^{2+} at 1305.91 eV. The occurrence of the C element (284.72 eV) can be attributed to the presence of hydrocarbons in air. The presence of O element (532.75 eV) should come from the hydrocarbon contaminations or MgO created by air oxidation [42]. As compared to the blank Mg, the O content on Mg-OH surface increased obviously while the content of C and Mg dropped substantially, suggesting that the sodium hydroxide treatment can produce a $Mg(OH)_2$ layer on the magnesium alloy surface. The existence of $Mg(OH)_2$ was also indicated by the O_{1s} single peak at 531.33 eV on the survey spectra. The high-resolution C_{1s} can be fitted into three peaks, the peak at 284.56 eV (C-C bond), the peaks at 285.25 and 287.91 eV, which can be attributed to the occurrence of $MgCO_3$. The formation of $MgCO_3$ may be the results of CO_2 in the air diffusing into the inner layer and reacting with the magnesium. The Mg_{1s} spectra may be divided into two peaks: the peak at 1304.65 eV corresponding to $MgCO_3$ and the peak at 1307.44 eV belonging to $Mg(OH)_2$ [43]. For Mg-HF, the elements of C, F, O, and Mg were detected on the survey spectra, and F_{1s} single peak at 690.79 eV were discovered in the high-resolution spectra, which corresponded to MgF_2, showing that MgF_2 was the primary component of the chemical conversion film. The O_{1s} peak can be attributed to the presence of a minor number of hydroxides, and the structure could be made of a mixture of $Mg(OH)_2$ and MgF_2 or a partial replacement of F^- in MgF_2 by OH^- [44]. The Mg_{1s} spectra can be divided into two peaks: one was the peak at 1309.27 eV corresponding to $Mg(OH)_2$, and the other peak at 1310.75 eV of MgF_2. The presence of new peaks of P_{2p} and Ba_{3d} on Mg-P surface showed that the oxide layer was successfully replaced by the phosphate coating. The O_{1s} spectra can be divided into two peaks: the peak at 532.82 eV arising from H_2O in the form of crystallization and the peak at 534.22 eV of P-OH group [45]. The high-resolution P_{2p} spectra is divided into five peaks: the peak of PO_4^{3-} at 133.32 eV, the peaks of HPO_4^{2-} at 134.03 eV and 134.73 eV, and the peaks of $H_2PO_4^-$ at 135.91 eV and 136.45 eV, demonstrating that P element was mainly existed in the surface as $Mg(H_2PO_4)_2$. Four peaks were used to match the Mg_{1s} spectra: MgO, $MgHPO_4$, $Mg_3(PO_4)_2$ and $Mg(H_2PO_4)_2$ at 1303.66 eV, 1304.25 eV, 1305.8 eV and 1306.83 eV, respectively [46]. As a result, $Mg_3(PO_4)_2$ and $Mg(H_2PO_4)_2$ were the main constitutes of the chemical conversion layer. The phytate-treated sample (Mg-PA) contained the elements O, C, P and Mg according to the survey spectra, and the presence of P (3.69%) demonstrated that the chemical conversion coating was successfully formed on the magnesium alloy surface. The O_{1s} spectra can be fitted into two peaks. One was the peak at 530.02 eV belonging to PO_4^{3-} and HPO_4^{2-}, proving that some of the O elements were from phytate. The other peak was the peak of magnesium oxides and hydroxides at 533.38 eV. The phytate radicals on oxides and CO_2 were represented by the fitted peaks at 284.77 and 288.74 eV on the C_{1s} spectra, respectively. P_{2p} displayed a single peak of PO_4^{3-} at 128.89 eV.

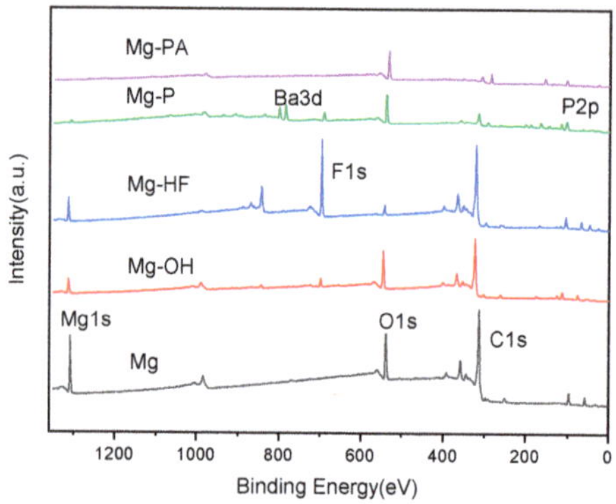

Figure 1. XPS survey spectra of the different samples.

Table 1. Surface atomic percentages of the different samples.

Samples	Mg	C	O	P	F	Ba
Mg	6.3	41.27	52.43	-	-	-
Mg-OH	2.99	29.99	67.02	-	-	-
Mg-HF	0.35	44.77	28.09	-	26.78	-
Mg-P	0.25	18.74	67.57	5.08	7.13	1.23
Mg-PA	0.24	43.89	52.18	3.69	-	-

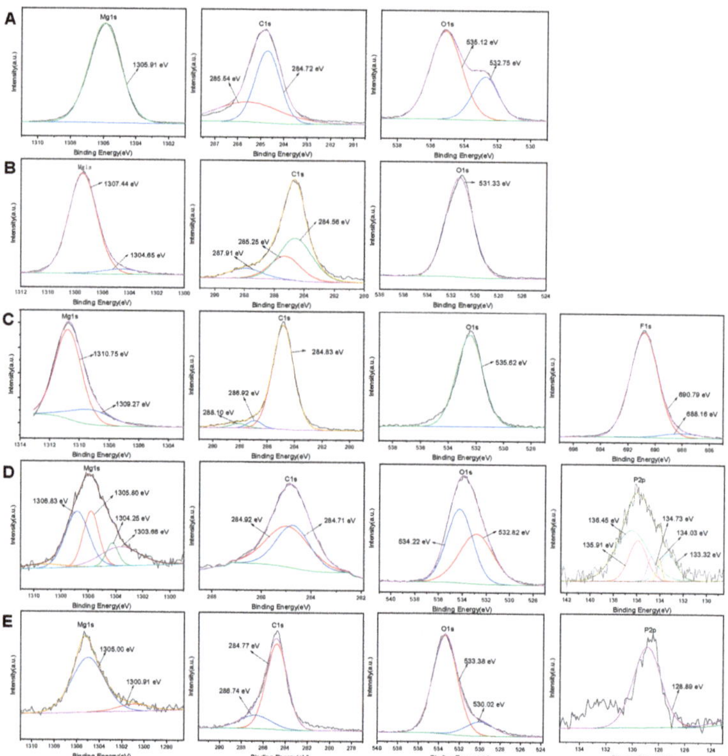

Figure 2. The high-resolution spectra of the main elements on the surfaces of different conversion films: (**A**) Mg; (**B**) Mg-OH; (**C**) Mg-HF; (**D**) Mg-P; (**E**) Mg-PA.

The surface morphologies of the different chemically treated magnesium alloys were observed by SEM, as illustrated in Figure 3. After polishing, the pristine magnesium alloy surface was relatively smooth. The surface of Mg-OH had a dense and homogenous $Mg(OH)_2$ layer, which slightly increased the surface roughness. The HF treatment can produce a dense and smooth coating on the surface, and there were also a few scattered irregular pores that may be due to the hydrogen generation and could be filled by MgF_2 and MgO particle precipitation [47]. A dense and lamellar-structured conversion film with visible cracks and pores can be observed on the Mg-P surface, which may result from the internal stresses during drying or when sodium fluoride is added, the entry of the F element results in the creation of hydrogen [48]. With the exception of a few small cracks, the surface chemical conversion film of the phytate-treated sample (Mg-PA) was intact and homogeneous, which was due to the fact that phytate contained twelve hydroxyls and six phosphate groups which can react with the dissolved metal ions of the magnesium alloy to form the water-insoluble magnesium phytate chelate. However, this reaction also produces a small amount of hydrogen, which passes through the surface layer to cause surface cracking [49]. In addition, it can be seen that both Mg-P and Mg-PA exhibited surface cracks that may be the result of substrate corrosion, and the presence of conversion layer flaws or poor crystallinity allows corrosion ions to diffuse into the surface of the substrate, resulting in the continuous anodic dissolution of the magnesium substrate [50]. The combination of XPS and SEM results demonstrated that the chemical conversion treatment can successfully create the different surface layers on the magnesium alloy surface. These coatings can isolate the substrate from the corrosive medium and stop reactive ions in solution from penetrating into the magnesium surface, which is helpful in reducing the corrosion rate of the magnesium substrate.

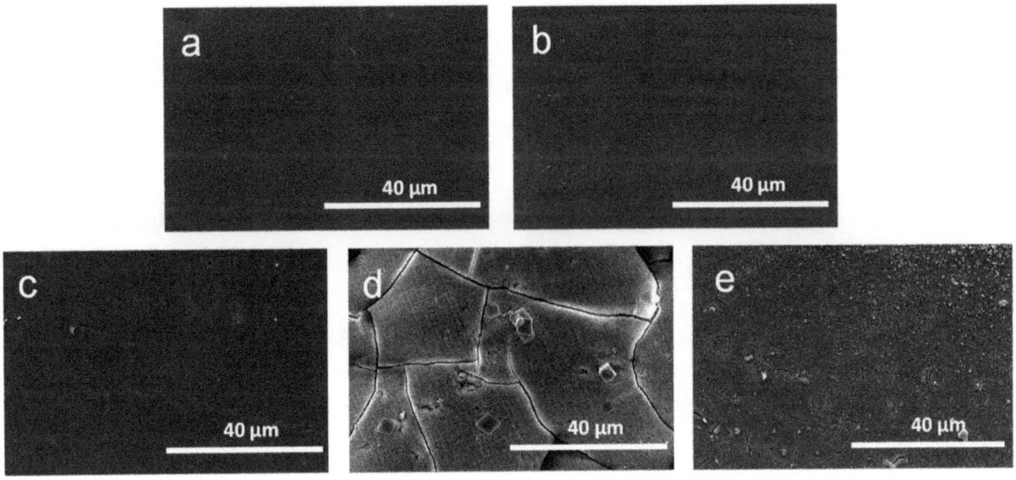

Figure 3. SEM images of the surface morphologies of the different samples: (**a**) Mg; (**b**) Mg-OH; (**c**) Mg-HF; (**d**) Mg-P; (**e**) Mg-PA.

Generally speaking, a good surface hydrophilicity is advantageous in improving the biocompatibility of the biomaterials because of the prolonged contact between the implant and the physiological environment in the body, as well as the substantial volume of water present in a human body. The surface hydrophilicity/hydrophobicity of bio-materials is often characterized by the water contact angle. In general, the hydrophilic surface is easy to wet when being in contact with water, and the water contact angle typically decreases as the surface roughness increases. Figure 4 displays the water contact angle results of the magnesium alloy treated by the different chemical treatments. As shown in Figure 4, the pristine magnesium alloy had a water contact angle of 56.3° and was relatively hydrophobic. The alkali heat treatment can produce a lot of hydroxyl groups on the magnesium alloy surface, which can combine with the water molecules to form hydrogen bonds and endow with a better hydrophilicity, resulting in a hydrophilicity of 12.5° for the Mg-OH. It has been demonstrated that the hydrophilic surface is useful for lowering the platelet adhesion and promoting cell adhesion and tissue growth [51]. The hydrophobic compound $Mg(OH)_xF_{2-x}$ on the surface of Mg-HF can cause an increase in the water contact angle to 103.2°, indicating that HF treatment can create a hydrophobic surface. Additionally, the increase in hydrophobicity also results in less surface adhesion between water droplets, which also raises the water contact angle. The water contact angles of Mg-P and Mg-PA were 12.7° and 19.7°, respectively. It was concluded that the presence of hydrophilic P-OH groups in $Mg_3(PO_4)_2$ and the hydroxyl groups in $Mg(OH)_2$ layer as well as the increased surface roughness can significantly reduce the water contact angle. Both the XPS results and the SEM images supported this conclusion.

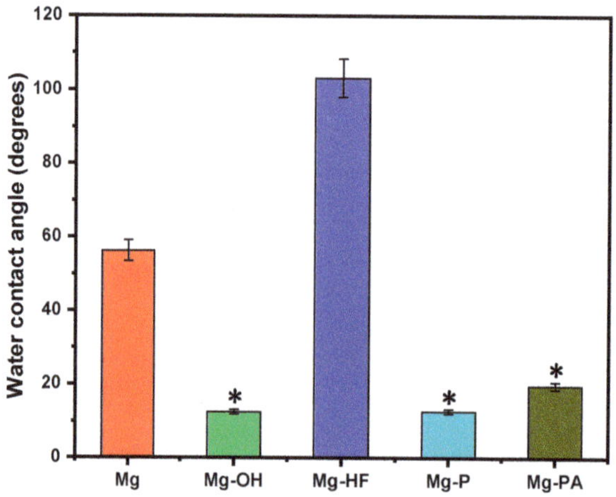

Figure 4. Water contact angles of the different modified magnesium alloy surfaces. The values are expressed as mean ± standard deviation (n = 3). The values of Mg-OH, Mg-P and Mg-PA are significantly lower ($p < 0.05$) than those of the pristine magnesium alloy in terms of hydrophilicity.

3.2. *Electrochemical Corrosion Behaviors*

The chemical structure, surface morphology, and microstructure of the surface chemical conversion layers are all strongly correlated with the corrosion resistance of the magnesium alloys [52]. In the present study, the potentiodynamic polarization curves were firstly measured to investigate the corrosion behaviors of the different magnesium alloys. In general, a higher corrosion potential (E_{corr}) often means a more stable thermodynamic state, while a lower corrosion current density (i_{corr}) means that the sample may corrode more slowly. Figure 5 shows the potentiodynamic polarization curves of the magnesium alloys with the different conversion layers, and Table 2 shows the corrosion potentials and corrosion current densities determined by the Tafel extrapolation method. It can be discovered that the corrosion potentials of all the chemical conversion layers were higher than that of the original magnesium alloy (−1.573 V), the corrosion current density (8.16×10^{-4} A·cm^{-2}) and the annual corrosion depth (18.5 mm/y) of the original magnesium alloy was also the highest, indicating that the chemical conversion treatment can improve the corrosion resistance of the magnesium alloy to varying degrees. This was due to the fact that the natural passivation layer on the pristine magnesium alloy was lacking density and susceptibility to corrosion in corrosive liquids. Additionally, the thermodynamic stability of the magnesium alloy increased thanks to the surface chemical treatment, which also slowed down the corrosion rate. As compared to the blank magnesium alloy, the corrosion current density of the chemical treated samples generally decreased by several orders of magnitude, indicating that these conversion coatings had excellent corrosion resistance when immersed in SBF solution. These coatings partially or completely isolate the magnesium substrate from the corrosion medium, preventing the corrosion medium from attacking the magnesium alloy substrate surface and thus lowering the corrosion rate. The MgF_2 coating on the Mg-HF surface is often created by the interatomic displacement of Mg and HF, and MgF_2 is insoluble in water, therefore it has a good shielding effect to slow down the corrosion rate. Additionally, Yan et al. [53] found that the presence of NaCl, $MgSO_4$, Na_2HPO_4 and $NaHCO_3$ in the simulated bodily fluids can effectively prevent the dissolving of MgF_2 coatings. The corrosion potential of Mg-P (−1.329 V) was higher than that of Mg-HF (−1.553 V), indicating that the phosphate film has a better thermodynamic stability compared to the MgF_2 film. Additionally, the laminar structure coating of the phosphate treatment contributed to increase the surface roughness, which can benefit the interfacial bonding of the magnesium matrix. The corrosion current density of Mg-PA (4.413×10^{-7} A·cm^{-2}) was three orders of magnitude lower than that of Mg, which resulted from the fact that phytic acid can chelate magnesium to form a stable phytic acid complex on the surface, and the excellent hydrophilicity was useful to improve the adhesion between the magnesium substrate and the coating, thereby reducing the corrosion rate. Among all samples, Mg-OH had the highest corrosion potential (−1.325 V), its corrosion current density (7.901×10^{-11} A·cm^{-2}) decreased by seven orders of magnitude compared to the pristine magnesium, and annual corrosion depth was only 1.79×10^{-6} mm/y, indicating that the chemical conversion layer after alkali heat treatment had the best anticorrosion performance. In addition, the alkali heat treatment also had the smooth surface and the smallest water contact angle, which can also contribute to the improvement of the corrosion resistance of the magnesium alloy.

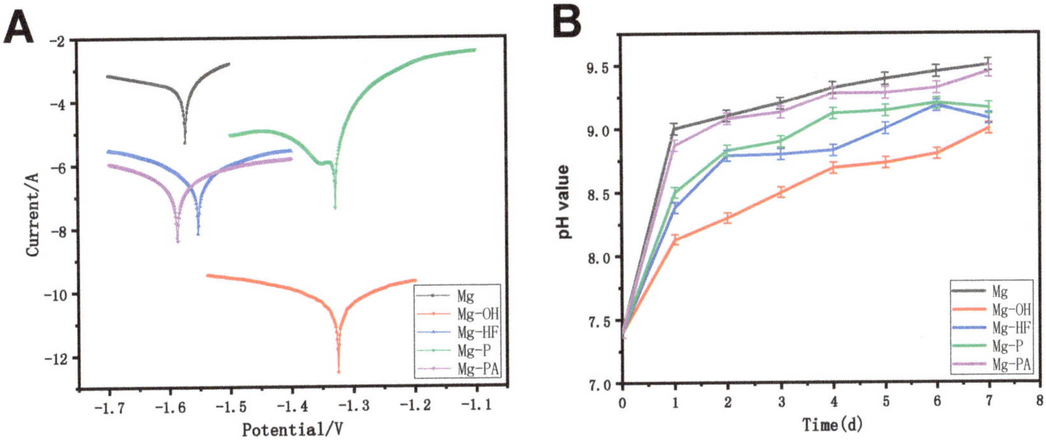

Figure 5. (**A**) The potentiodynamic polarization curves of the different samples. (**B**) The pH curves of the different samples immersed in SBF solution for 7 days.

Table 2. Corrosion potential and corrosion current density of the different samples.

Samples	E_{corr}/(V)	i_{corr}/(A·cm^{-2})	d/(mm/y)
Mg	−1.573	8.16×10^{-4}	18.5
Mg-OH	−1.325	7.901×10^{-11}	1.79×10^{-6}
Mg-HF	−1.553	3.000×10^{-6}	6.79×10^{-2}
Mg-P	−1.329	1.798×10^{-6}	4.07×10^{-2}
Mg-PA	−1.587	4.413×10^{-7}	9.98×10^{-3}

The physiological environment of the human body has a weak alkaline pH of roughly 7.4. According to the corrosion mechanism of the magnesium alloy, corrosion degradation results in a significant amount of corrosion products, including hydrogen gas and hydroxide, which raises the pH value and thus may cause a series of adverse physiological reactions. Figure 5B displays the pH change curves for the different samples dipped in SBF solution for seven days. It is obvious that the pH value for the magnesium alloys both before and after the chemical treatments increased with the increase in the immersion time. As the surface conversion film started to dissolve at the initial stage of the immersion, the structure may become loose, the corrosion medium permeated the coating, and subsequently corroded the magnesium substrate. The corrosion of the magnesium alloy was often brought on by an oxidation-reduction reaction in which the anode released Mg^{2+} and the cathode precipitated hydrogen [54]. As water gains electrons, gaseous hydrogen and hydroxide ions were produced. The pH value of each sample both exhibited a rising trend due to the formation of OH^-, and the corrosion products that were gradually accumulating on the magnesium alloy surface caused the degradation rate decreased generally and the curve rose slowly. The pH value of the pristine magnesium alloy grew the fastest, rising to 9.5 in just 7 days, suggesting that the unmodified magnesium alloy had the lowest anticorrosion performance and the quickest degradation rate. The other modified samples showed a slower pH rising trend, indicating that the surface chemical conversion layer could prevent the corrosion of the magnesium alloy and then reduce the degradation rate to different degrees. From 2 days to 4 days, the rising trend of the curve slowed down because ${SO_4}^{2-}$ and ${HCO_3}^{2-}$ in the simulated body fluid can inhibit the corrosion to some extent. Song et al. [55] also found that sulfate or bicarbonate replaced chloride ions adsorbed on the magnesium surface and passivated the active sites on the magnesium surface, resulting in a reduction in the corrosion rate. As the immersion time increased, the corrosion products increasingly accumulated and made it harder for the corrosion medium to contact with the magnesium substrate. This resulted in a slower degradation rate, which was the reason for the stable pH curve of Mg-HF. After 6 days, the dissolving of the soluble $MgHPO_4$ on the magnesium alloy surface and the reformation of the less soluble $Mg(OH)_2$ product can substantially slow down the degradation rate of the coating, which may be the reason of the drop in pH value of Mg-HF and Mg-P. The pH value of the alkali heat-treated samples was consistently lower than other coated magnesium alloys, showing that the Mg-OH samples was the most corrosion-resistant of all the coated samples. Additionally, the SEM microscopic images indicated that there were the significant cracks in the phosphate and phytate treated surfaces, indicating that these conversion layers had weak corrosion resistance. In contrast, the sodium hydroxide-treated sample surface was the densest.

Magnesium alloys treated by the different chemical treatment were immersed into SBF solution for 1, 3, 7 and 14 days in order to further characterize the corrosion resistance. The surface corrosion morphologies and elemental contents of the samples after the corrosion were then observed and analyzed by SEM and EDS, respectively, and the results are shown in Figure 6 and Table 3. After soaking for one day, tiny fractures started to emerge on the original magnesium alloy surface, demonstrating that not only the pristine magnesium alloy had poor capacity to withstand the corrosion but also the naturally occurring oxide layer had the inability to withstand the aggressive ions in the solution. Small micro-cracks that formed on Mg-P and Mg-PA surfaces were believed to be the result of local substrate corrosion brought on by chloride ion penetration during the immersion. The Mg-OH and Mg-HF surfaces were no cracks, it was considered that the dense $Mg(OH)_2$ passivation layer and MgF_2 coating can reduce the etching of corrosive ions and thus prevent the substrate from corroding. After three days, Mg, Mg-P, and Mg-PA sample surface fractures grew and deepened, and Mg-HF and Mg-OH sample cracks first appeared, indicating that the modified surface was initially quite protective but gradually loses the protection over time. However, the cracked coating was still attached to the magnesium surface, and it still provided a certain degree of protection. The corrosion conditions for all samples became worse with longer immersion times of 7 days. Shrinkage and dehydration of corrosion products may have an impact on the appearance of wide cracks [41]. Mg surface corrosion was particularly severe, and the appearance of many corrosion products was brought on by the reaction of corrosion ions (Cl^-) with the initial layer of $Mg(OH)_2$ on the surface to produce highly water-soluble $MgCl_2$ and release hydroxyl groups. Additionally, Barajas et al. [56] found that mechanical stresses that might cause cracking can be produced by the slow accumulation and expansion of corrosion products on the magnesium alloy surfaces. On the Mg-OH surface, there were small fractures to be observed, indicating that the rough surface of this sample increased the contact area with the corrosive medium and the protective property of the chemical conversion layer became gradually weakening. In 14 days, the coating protection was decreased further. Corrosion pits appeared on the surface of Mg and Mg-P, and the corrosive medium readily contacted the surfaces of the magnesium matrix, increasing the susceptibility to corrosion and encouraging the accumulation and deterioration of the corrosion products. This was most likely because the magnesium matrix acted as an anode with the second phase as a local micro cathode for internal galvanic coupling corrosion, which speeded up its corrosion [57]. The EDS findings demonstrated that the Mg surface corrosion was the most severe, with an increase in the contents of O and Ca and a decrease in the contents of Mg, therefore the corrosion products were probably calcium-rich compounds. Due to surface corrosion and coating deterioration, which allowed magnesium ions from the solution to enter the coating and adsorb to the surface, the Ca content of Mg-P dropped to zero and the Mg content increased. The Mg-HF and Mg-PA coatings had significant amounts of the corrosion products deposited on their surface, likely as a result of the prolonged immersion time. At the end of the reaction, the corrosion products were deposited on the sample surface. Mg, O, P, and Ca were abundant on the surfaces of Mg-HF and Mg-PA, and hydroxides and phosphates, which were corrosion products, may also be present. However, there were only reticulated cracks on the Mg-OH surface, there was no clear surface corrosion, and the EDS data showed no change in the content of Mg, O and C, demonstrating the superiority of the $Mg(OH)_2$ conversion layer in enhancing the magnesium alloy corrosion resistance.

3.3. Protein Adsorption

There are three main proteins in human plasma, including albumin, globulin, and fibrinogen. Albumin is a mild, non-reactive protein which can maintain plasma osmolality and ensure normal transport of lipids and steroid hormones. The albumin layer adsorbed on the surface also inhibits the thrombin produced by contact between implants and blood substances, resulting in an antithrombotic effect [58]. Fibrinogen participates in the coagulation and thrombosis processes, and its interactions with platelets can have negative consequences on blood. The amounts of the adsorbed albumin and fibrinogen are shown in Figure 7. It can be seen that Mg-OH and Mg-P had the most albumin adsorbed on them. Albumin can use conformational changes to gain entropy, which enables more albumin to be adsorbed on the hydrophilic surface and delays thrombus formation [59]. On the other hand, albumin molecules are negatively charged and it can interact with Mg^{2+} to adsorb onto the magnesium surface, which may inhibit magnesium anodic dissolution. It has been reported that the albumin adsorption can act as a good shield to block the reactive sites on the magnesium surface, and thus inhibit the penetration of corrosive ions to some extent, finally leading to the improved corrosion resistance [60]. Since proteins have a greater propensity to connect to hydrophobic surfaces and establish stronger interactions, this may explain why Mg-HF absorbed more albumin Figure 7B shows that Mg-HF adsorbed the most fibrinogen because the hydrophobic surface can interact with the hydrophobic fibrinogen. The existence of a powerful attraction between fibrinogen molecules and hydrophobic surfaces was confirmed by Paul et al. [61]. As compared to Mg, both Mg-P and Mg-PA absorbed significantly more albumin and fibrinogen, demonstrating that the modified surface can simultaneously enhance albumin and fibrinogen adsorption. It was noteworthy that Mg-OH had the lowest fibrinogen uptake capacity, along with the improved albumin adsorption, suggesting that the NaOH-treated magnesium alloy had the ability to selectively adsorb albumin and may therefore have better hemocompatibility.

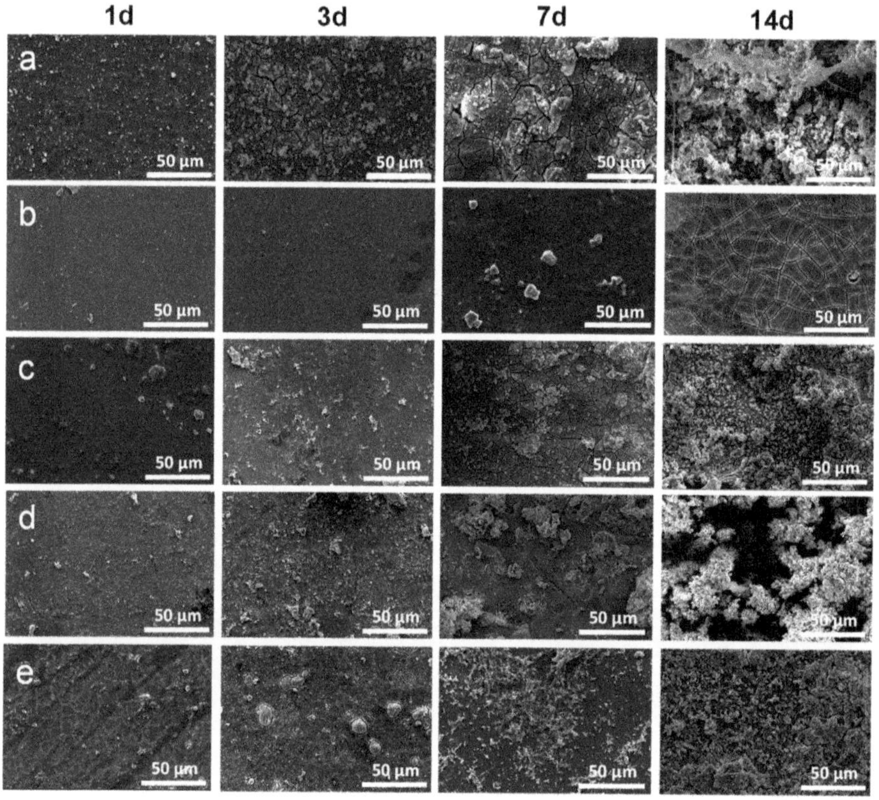

Figure 6. Typical SEM images of the different samples immersed in SBF solution for 1, 3, 7 and 14 days: (**a**) Mg; (**b**) Mg-OH; (**c**) Mg-HF; (**d**) Mg-P; (**e**) Mg-PA.

Table 3. Surface element contents of the different samples after 14 d immersion in SBF.

Samples	Mg	C	O	P	F	Ca
Mg	7.23	-	66.12	-	-	26.65
Mg-OH	8.83	51.95	39.22	-	-	-
Mg-HF	4.24	25.64	58.21	7.22	3.52	4.68
Mg-P	23.28	24.68	52.04	-	-	-
Mg-PA	9.98	30.87	50.06	3.82	-	5.27

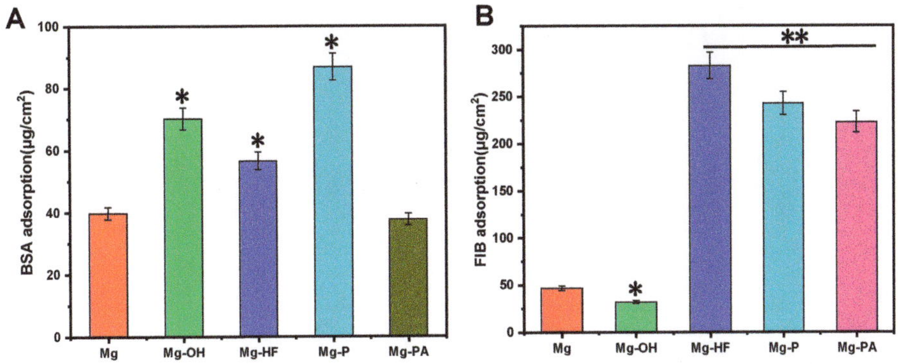

Figure 7. The amounts of albumin (**A**) and fibrinogen (**B**) adsorbed on the different surfaces. Data for each sample were taken from three parallel samples, and are expressed as mean ± SD. The samples of Mg-OH, Mg-HF and Mg-P showed significantly differences (* $p < 0.05$) in promoting albumin adsorption compared with the pristine magnesium alloy. The samples of Mg-HF, Mg-P and Mg-PA showed significant differences (** $p < 0.05$) in fibrinogen adsorption compared with the pristine magnesium alloy and Mg-OH.

3.4. *Blood Compatibility*

The implant that comes into contact with the blood must be highly hemocompatible; otherwise, they have the risk of causing a variety of unfavorable reactions, including coagulation, complement activation, platelet activation and leukocyte destruction, which can result in thrombus formation and hemolysis. The hemolysis rate is a crucial factor in determining the hemocompatibility of biomaterials. Better hemocompatibility is indicated by the lower hemolysis. Hemolysis is often influenced by a variety of variables, including chemical compounds, pH levels, and metal ion concentrations, etc. The maximum acceptable hemolysis rate for the blood-contact biomaterials is 5%, in accordance with the ISO 10993-4:2002 standard. Figure 8A shows the results of the hemolysis rates of the different samples. The highest rate of hemolysis can be found on the pristine magnesium alloy, indicating that it may result in severe hemolysis when contacted with human blood. The pristine magnesium can corrode in blood to release a large amount of Mg^{2+}, which could result in high osmotic pressure of blood cells and therefore lead to swelling of erythrocytes until membrane rupture, finally leading to the higher hemolysis rate [62]. Additionally, the degradation of the magnesium alloy resulted in the production of OH^-, which raised the pH of the blood. The alkaline microenvironment also damages erythrocytes by inducing instability in their membranes [62], which also contributed to severe hemolysis. The surface-modified samples showed a considerable reduction in the rate of hemolysis and all met the hemolysis requirements of the blood-contacting bio-materials, showing that these chemical surface treatments do not seriously harm red blood cells and can effectively prevent the appearance of hemolysis. Due to the better corrosion resistance of $Mg(OH)_2$ layer after NaOH treatment, the decreased release of Mg^{2+} can significantly lower the hemolysis rate when the magnesium matrix came into contact with the blood. The hydrophobic of Mg-HF had water-resistant property, which can slow down the degradation rate of the magnesium alloy and therefore reduce the hemolysis rate; on the other hand, the repulsion to blood resulted in low red cells adhesion. The phosphate coating on Mg-P can prevent the Cl^- adsorption by electrostatic repulsion because of the presence of the negatively charged groups (PO_4^{3-}). It can obviously hamper the magnesium matrix from harming the red blood cells and hence preventing the thrombosis [8]. The lowest hemolysis rate was found in the Mg-PA samples, indicating that the phytate macromolecules were chemically bonded to the magnesium ions and other cations to form dense chelates that can prevent direct contact between the magnesium matrix and red blood cells, resulting in the least damage to the red blood cells and good anti-hemolytic properties [63].

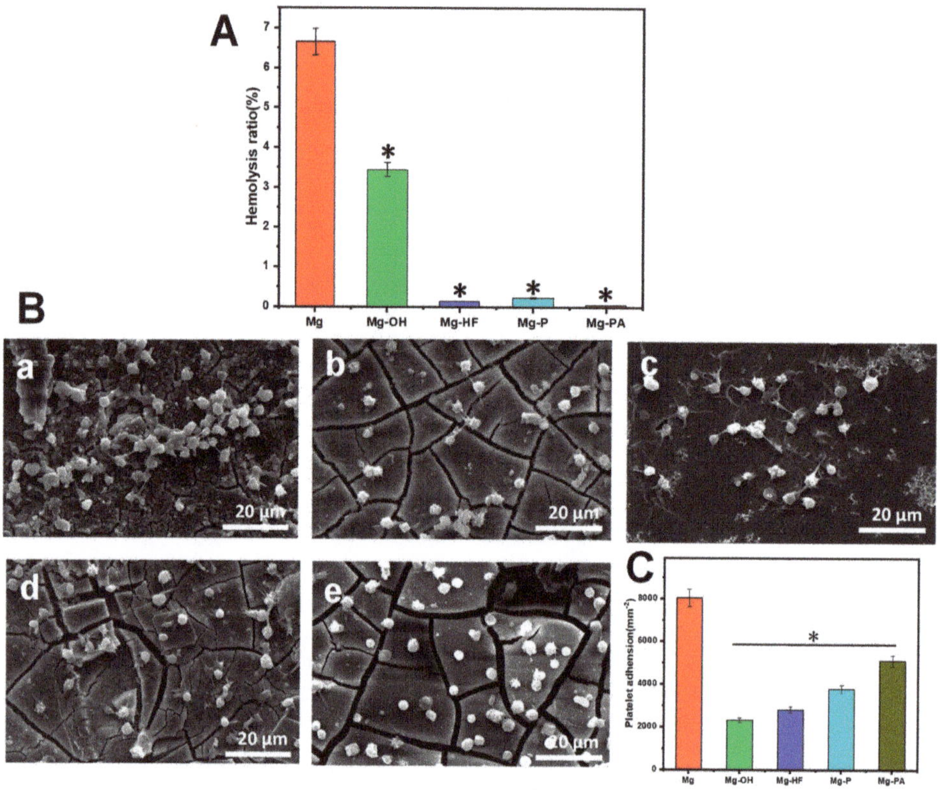

Figure 8. (**A**) Hemolysis rates of the different samples. (**B**) Typical SEM images of platelet adhesion on the surface of different samples: (**a**) Mg; (**b**) Mg-OH; (**c**) Mg-HF; (**d**) Mg-P; (**e**) Mg-PA. (**C**) The number of the platelets adhered on the different samples. For hemolysis rate and platelet number, the data are presented as mean ± SD of three parallel samples. The values of the modified samples are significantly lower (* $p < 0.05$) than that of the pristine magnesium alloy.

Platelet adhesion and activation play a vital role in coagulation responses. When a poorly blood-compatible biomaterial is exposed to blood, platelets can attach to it and undergo shape changes, aggregation, and activation, leading to thrombus formation [64]. The quantity of platelets attached to the surface, morphological changes, and the expression of particular proteins on the platelet membrane can all be used to investigate the degree of the platelet behaviors [65]. In the present study, scanning electron microscopy (SEM) images were taken to reveal the quantity and the morphological changes in the attached platelets on the different surfaces, and the results are shown in Figure 8B,C. It can be clearly seen that the highest number of platelets aggregated on the pristine magnesium alloy surface. It was considered that the most severe surface corrosion and surface hydrophobicity may promote platelet adhesion and activation, and the high pH value caused by an excessively rapid corrosion rate also promoted the platelet activation. This suggested that the pure magnesium alloy had poor blood compatibility, by contrast, there was an obvious reduction in the number of platelets adhered on the chemical treated magnesium alloy surfaces (Figure 8C). The Mg-OH displayed the lowest number of the platelets attached on the surface, and this was because that the alkali heat treatment had created a chemical passivation layer with active hydroxyl groups on the surface, which can improve the hydrophilicity and corrosion resistance and further inhibit platelet adhesion and activation. In addition, it can be seen that almost all of the platelets on Mg-OH were still present and showed no inclination to become activated, which could be conducive to prevent thrombus formation after the implantation. The Mg-HF and Mg-P samples had comparatively less platelets adhesion and activation as compared to the blank magnesium alloy, indicating that these two chemical treatments can enhance the anticoagulation of the magnesium alloy to some degree. It was concluded that the improved corrosion resistance can significantly prevent the release of Mg^{2+} and local alkalization, which contributed to enhance the anticoagulation. In addition, the magnesium alloy surface was negatively charged after the phosphate treatment, which can also prevent the adhesion of the negative-charged platelets [26]. For the Mg-P samples, phosphate treatment can result in low platelet adhesion and diffusion to create a good blood contact surface and avoid excessive platelet aggregation. Additionally, the generation of hydrogen and corrosion products has a hindering impact on the approaching platelets [36]. In addition, the sample surface showed a small and thin pseudopod on Mg-HF and Mg-P, indicating a slight activation. Since fibrinogen can be attracted to the surface of Mg-PA, which can promote thrombogenic platelet adhesion and aggregation, as evidenced by the increased platelet count and partial activation with pseudopod morphology.

3.5. *Endothelial Cell Behaviors*

3.5.1. Cell Adhesion

One of the most advantageous characteristics of the vascular implants is their propensity to promote the growth of endothelial cells (ECs) on the surface since ECs have natural biocompatibility and constitute the inner wall of the artery. Biomaterials' biocompatibility is demonstrated by the adherence and growth of cells on them [66]. Typically, when cells come into touch with biomaterials, they could alter their shape in order to integrate the materials and cells. Zhang et al. [67] reported the cell adhesion on the surface of magnesium alloy cultured for 6 and 24 h and found that the number of adherent cells was lower at 6 h, while adhesion would be better at 24 h. Therefore, we began to record the cell adhesion from 24 h. The fluorescence staining images of the endothelial cell adhered on the different surfaces after 1 and 2 days of incubation are shown in Figure 9. It can be seen that the cells can attach and grow on all samples, but the original magnesium alloy surface mainly had the rounded endothelial cells because of its limited corrosion resistance. Large amounts of the hydroxide might cause partial alkalinization of the implant, and the formation of hydrogen gas bubbles on the surface of the magnesium matrix may prevent cell adhesion and growth [68]. After 2 days, the number of cells dramatically increased for the modified samples and covered the majority of the surface. It was considered that the surface coating can slow down the degradation rate and increase the number of cell adhesion. On the other hand, a good surface wettability can create a favorable weak alkaline milieu for cell growth and boost cell adhesion and proliferation, therefore, the number of cells on the Mg-OH surface grew better. In addition, the moderately wetted surface induces the adsorption and activation of proteins, especially mucilage proteins, which promotes cell adhesion and improves surface bioactivity [69]. The MgF_2 coating on Mg-HF had the capacity to prevent the excessive release of Mg^{2+} and can enhance cell adherence and proliferation on the surface by a good survival interface [70]. In general, the hydrophilic surface facilitates cell adhesion, migration, and proliferation. Cell adhesion and growth for Mg-P and Mg-PA were positively influenced by the hydrophilicity, coupled with the presence of hydrophilic phosphate groups, it can improve the cell adhesion and differentiation. However, if the immersion period was too long. It may corrode readily and have an adverse impact on cell proliferation. When compared to the pure magnesium, all the modified samples displayed a considerable increase in cell density, proving that these coatings can promote endothelial cell adhesion and growth.

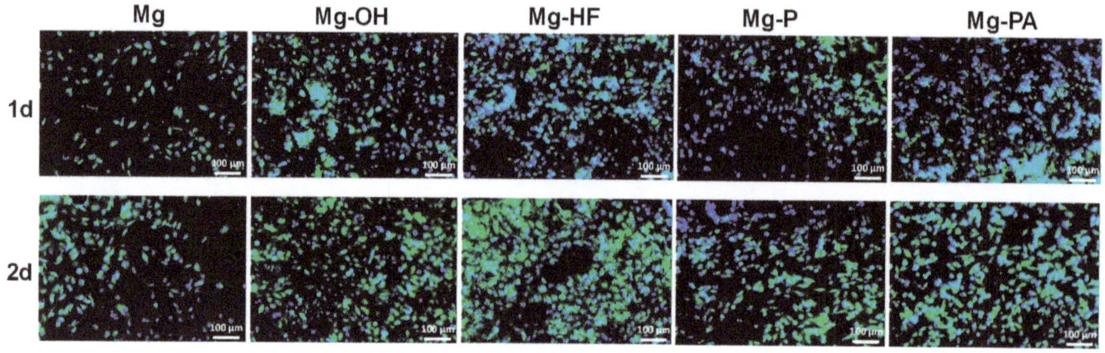

Figure 9. The typical fluorescence images of cells attached to the surface of different samples cultured for 1d and 2d.

3.5.2. Cell Proliferation

CCK-8 is typically used to the evaluate endothelial cell survival and proliferation. Figure 10 displays the CCK-8 values of the endothelial cells at 1 and 2 days of growth with the different modified magnesium alloys. It was obvious that the pristine magnesium had the lowest CCK-8 value. On the one hand, it was possible that the original magnesium alloy surface corrosion released a large amount of metal ions and hydroxides that were harmful to cell growth. On the other hand, the modification of the magnesium alloy surface can create a chemical conversion layer that was resistant to corrosion, which can achieve the goal of reducing the corrosion rate and reducing the release of metal ions, effectively promoting cell proliferation. The highest CCK-8 value can be detected for Mg-OH in 1 day, suggesting that the alkali heat treatment not only introduced the hydrophilic groups to the magnesium alloy surface, but can also promote cell growth and proliferation. Compared to Mg-P and Mg-PA, Mg-HF had a greater CCK-8 value because of the hydrophobic surface and roughness of the MgF_2 layer created by the fluorination treatment, which may increase the attraction to sticky proteins and aid in the stimulation of cell adhesion and proliferation. At 2 days, Mg, Mg-P, and Mg-PA had increased cell proliferation, whereas Mg-OH and Mg-HF had decreased cell proliferation. Nevertheless, Mg-OH has the highest CCK-8 value. It was possible that the force necessary for cell adhesion decreased linearly with the increase in the surface wettability, and once the critical shear stress was reached, the cells were unable to adhere to the surface, impairing endothelial cells ability to grow and proliferate normally. This could explain why the proliferation capacity of the Mg-OH sample decreased [71]. Regarding the Mg-HF sample, it was possible that the immersion time was too long, leading to the corrosion of the MgF_2 layer, which was not positive for cell survival. Alternatively, the cells may fall off from the surface due to the production of hydrogen and corrosion products, making it difficult for the cells to attach and proliferate [72]. Generally speaking, cells tend to prefer hydrophilic surfaces for adhesion, spreading, and growth. Since Mg-P and Mg-PA surfaces were rougher than Mg-HF, there was more surface area available for cell adhesion and proliferation, leading to higher CCK-8 values than Mg-HF. According to the results of cell adhesion and proliferation, the Mg-OH surface had good bioactivity and biocompatibility, which created an ideal environment for cell adhesion and proliferation, suggesting that the NaOH-treated samples had good properties of promoting endothelial cell adhesion and growth.

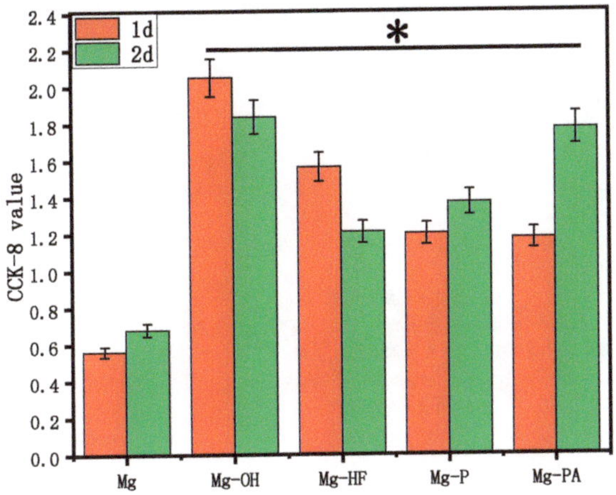

Figure 10. CCK-8 values of the endothelial cells cultured on different sample surfaces for 1d and 2d. Data were obtained from three parallel samples and are expressed as mean ± SD. * $p < 0.05$ indicated the statistical differences compared with Mg.

4. Conclusions

The different chemical conversion layers were successfully produced on the magnesium alloy surface by sodium hydroxide (NaOH), hydrofluoric acid (HF), phosphoric acid (H_3PO_4) and phytic acid ($C_6H_{18}O_{24}P_6$) treatment, and the as-prepared chemical conversion layers showed the different surface chemical structures and surface morphologies, leading to significant differences in their wettability. The magnesium alloy treated by sodium hydroxide had the best hydrophilicity and a dense structure. Based on the chemical structures of the conversion layers, the different conversion layers exhibited various behaviors of electrochemical corrosion degradation and protein adsorption. In contrast, the magnesium alloy after sodium hydroxide treatment exhibited the best corrosion resistance, the least degree of corrosive degradation, and the capacity to preferentially absorb albumin, which also led to superior anticoagulant characteristics and the ability to withstand corrosion. Although all of the chemically treated samples could enhance the endothelial cell adhesion and proliferation to some degree, the good endothelial cell adhesion and proliferation of the sodium hydroxide-treated magnesium alloy were demonstrated by the comparatively high number of cell adhesions and the highest CCK-8 values. Therefore, among all of the chemical conversion treatments, the surface sodium hydroxide treatment produced a dense and excellent hydrophilic coating that exhibited excellent overall performance and can be applied to magnesium alloy vascular stent materials to improve their biocompatibility and corrosion resistance. In addition, sodium hydroxide treatment can also introduce a large number of hydroxyl groups on the magnesium alloy surface, which makes it easier to further modify the surface by biofunctionalization and thus further enhance the corrosion resistance and biocompatibility of the magnesium alloy.

Author Contributions: Conceptualization, L.M. and C.P.; methodology, L.M.; software, L.M., L.L. and F.G.; validation, F.G.; formal analysis, L.M. and L.L.; investigation, L.M. and Q.H.; resources, Y.C. and C.P.; data curation, X.L.; writing—original draft preparation, L.M. and X.L.; writing—review and editing, J.C., Q.Z. and C.P.; visualization, Q.H. and J.C.; supervision, Y.C. and C.P.; funding acquisition, C.P. All authors have read and agreed to the published version of the manuscript.

Funding: This research was funded by the National Natural Science Foundation of China (31870952), Natural Science Foundation of Jiangsu Province of China (BK20181480) and Natural Science Foundation of Huaiyin Institute of Technology(22HGZ003).

Data Availability Statement: Not applicable.

Conflicts of Interest: The authors declare no conflict of interest.

References

1. Foerst, J.R.; Gruenheck, F.; Kelm, M.; Vorpahl, M. Coronary Artery Stents: Approved First-Generation Drug-Eluting Stents. In *Drug-Eluting Stents*; Future Medicine Ltd.: London, UK, 2012; pp. 18–30. ISBN 978-1-78084-056-7.
2. Saito, S. New Horizon of Bioabsorbable Stent. *Cathet. Cardiovasc. Intervent.* **2005**, *66*, 595–596. [CrossRef] [PubMed]
3. Byrne, R.A.; Sarafoff, N.; Kastrati, A.; Schömig, A. Drug-Eluting Stents in Percutaneous Coronary Intervention: A Benefit-Risk Assessment. *Drug Saf.* **2009**, *32*, 749–770. [CrossRef] [PubMed]
4. Erne, P.; Schier, M.; Resink, T.J. The Road to Bioabsorbable Stents: Reaching Clinical Reality? *Cardiovasc. Interv. Radiol.* **2006**, *29*, 11–16. [CrossRef] [PubMed]
5. Schömig, A.; Dibra, A.; Windecker, S.; Mehilli, J.; Suárez de Lezo, J.; Kaiser, C.; Park, S.-J.; Goy, J.-J.; Lee, J.-H.; Di Lorenzo, E.; et al. A Meta-Analysis of 16 Randomized Trials of Sirolimus-Eluting Stents Versus Paclitaxel-Eluting Stents in Patients with Coronary Artery Disease. *J. Am. Coll. Cardiol.* **2007**, *50*, 1373–1380. [CrossRef] [PubMed]
6. Zhang, E.; Shen, F. Blood Compatibility of a Ferulic Acid (FA)-Eluting PHBHHx System for Biodegradable Magnesium Stent Application. *Mater. Sci. Eng. C* **2015**, *52*, 37–45. [CrossRef]
7. Chen, Y.; Dou, J.; Yu, H.; Chen, C. Degradable Magnesium-Based Alloys for Biomedical Applications: The Role of Critical Alloying Elements. *J. Biomater. Appl.* **2019**, *33*, 1348–1372. [CrossRef] [PubMed]
8. Zhang, Z.-Q.; Wang, L.; Zeng, M.-Q.; Zeng, R.-C.; Kannan, M.B.; Lin, C.-G.; Zheng, Y.-F. Biodegradation Behavior of Micro-Arc Oxidation Coating on Magnesium Alloy-from a Protein Perspective. *Bioact. Mater.* **2020**, *5*, 398–409. [CrossRef]
9. Yan, W.; Lian, Y.-J.; Zhang, Z.-Y.; Zeng, M.-Q.; Zhang, Z.-Q.; Yin, Z.-Z.; Cui, L.-Y.; Zeng, R.-C. In Vitro Degradation of Pure Magnesium—the Synergetic Influences of Glucose and Albumin. *Bioact. Mater.* **2020**, *5*, 318–333. [CrossRef]
10. Staiger, M.P.; Pietak, A.M.; Huadmai, J.; Dias, G. Magnesium and Its Alloys as Orthopedic Biomaterials: A Review. *Biomaterials* **2006**, *27*, 1728–1734. [CrossRef]
11. Ford, E.S.; Mokdad, A.H. Dietary Magnesium Intake in a National Sample of U.S. Adults. *J. Nutr.* **2003**, *133*, 2879–2882. [CrossRef]
12. Zhang, Y.; Xu, J.; Ruan, Y.C.; Yu, M.K.; O'Laughlin, M.; Wise, H.; Chen, D.; Tian, L.; Shi, D.; Wang, J.; et al. Implant-Derived Magnesium Induces Local Neuronal Production of CGRP to Improve Bone-Fracture Healing in Rats. *Nat. Med.* **2016**, *22*, 1160–1169. [CrossRef] [PubMed]
13. Shi, Z.; Atrens, A. An Innovative Specimen Configuration for the Study of Mg Corrosion. *Corros. Sci.* **2011**, *53*, 226–246. [CrossRef]
14. Zeng, R.; Dietzel, W.; Witte, F.; Hort, N.; Blawert, C. Progress and Challenge for Magnesium Alloys as Biomaterials. *Adv. Eng. Mater.* **2008**, *10*, B3–B14. [CrossRef]
15. Esmaily, M.; Svensson, J.E.; Fajardo, S.; Birbilis, N.; Frankel, G.S.; Virtanen, S.; Arrabal, R.; Thomas, S.; Johansson, L.G. Fundamentals and Advances in Magnesium Alloy Corrosion. *Prog. Mater. Sci.* **2017**, *89*, 92–193. [CrossRef]
16. Khang, D.; Lu, J.; Yao, C.; Haberstroh, K.M.; Webster, T.J. The Role of Nanometer and Sub-Micron Surface Features on Vascular and Bone Cell Adhesion on Titanium. *Biomaterials* **2008**, *29*, 970–983. [CrossRef] [PubMed]
17. Ahuja, N.; Batra, U.; Kumar, K. Experimental Investigation and Optimization of Wire Electrical Discharge Machining for Surface Characteristics and Corrosion Rate of Biodegradable Mg Alloy. *J. Mater. Eng. Perform.* **2020**, *29*, 4117–4129. [CrossRef]
18. Sharma, G.; Kumar, K.; Satsangi, P.S.; Sharma, N. Surface Modification of Biodegradable Mg-4Zn Alloy Using PMEDM: An Experimental Investigation, Optimization and Corrosion Analysis. *IRBM* **2021**, *43*, S1959031821000245. [CrossRef]
19. Li, D.; Yuan, Q.; Yu, K.; Xiao, T.; Liu, L.; Dai, Y.; Xiong, L.; Zhang, B.; Li, A. Mg–Zn–Mn Alloy Extract Induces the Angiogenesis of Human Umbilical Vein Endothelial Cells via FGF/FGFR Signaling Pathway. *Biochem. Biophys. Res. Commun.* **2019**, *514*, 618–624. [CrossRef]
20. Xiong, X.; Yang, Y.; Li, J.; Li, M.; Peng, J.; Wen, C.; Peng, X. Research on the Microstructure and Properties of a Multi-Pass Friction Stir Processed 6061Al Coating for AZ31 Mg Alloy. *J. Magnes. Alloy.* **2019**, *7*, 696–706. [CrossRef]
21. Mhaede, M.; Pastorek, F.; Hadzima, B. Influence of Shot Peening on Corrosion Properties of Biocompatible Magnesium Alloy AZ31 Coated by Dicalcium Phosphate Dihydrate (DCPD). *Mater. Sci. Eng. C* **2014**, *39*, 330–335. [CrossRef]
22. Liu, J.; Zheng, Y.; Bi, Y.; Li, Y.; Zheng, Y. Improved Cytocompatibility of Mg-1Ca Alloy Modified by Zn Ion Implantation and Deposition. *Mater. Lett.* **2017**, *205*, 87–89. [CrossRef]
23. Zeng, R.-C.; Jiang, K.; Li, S.-Q.; Zhang, F.; Cui, H.-Z.; Han, E.-H. Mechanical and Corrosion Properties of Al/Ti Film on Magnesium Alloy AZ31B. *Front. Mater. Sci.* **2015**, *9*, 66–76. [CrossRef]
24. Cui, X.; Jin, G.; Li, Q.; Yang, Y.; Li, Y.; Wang, F. Electroless Ni–P Plating with a Phytic Acid Pretreatment on AZ91D Magnesium Alloy. *Mater. Chem. Phys.* **2010**, *121*, 308–313. [CrossRef]
25. Guo, Y.; Zhang, Y.; Li, Z.; Wei, S.; Zhang, T.; Yang, L.; Liu, S. Microstructure and Properties of In-Situ Synthesized ZrC-Al3Zr Reinforced Composite Coating on AZ91D Magnesium Alloy by Laser Cladding. *Surf. Coat. Technol.* **2018**, *334*, 471–478. [CrossRef]
26. Mao, L.; Yuan, G.; Niu, J.; Zong, Y.; Ding, W. In Vitro Degradation Behavior and Biocompatibility of Mg–Nd–Zn–Zr Alloy by Hydrofluoric Acid Treatment. *Mater. Sci. Eng. C* **2013**, *33*, 242–250. [CrossRef]
27. Xue, D.; Yun, Y.; Schulz, M.J.; Shanov, V. Corrosion Protection of Biodegradable Magnesium Implants Using Anodization. *Mater. Sci. Eng. C* **2011**, *31*, 215–223. [CrossRef]

28. Yang, X.; Li, M.; Lin, X.; Tan, L.; Lan, G.; Li, L.; Yin, Q.; Xia, H.; Zhang, Y.; Yang, K. Enhanced In Vitro Biocompatibility/Bioactivity of Biodegradable Mg–Zn–Zr Alloy by Micro-Arc Oxidation Coating Contained Mg_2SiO_4. *Surf. Coat. Technol.* **2013**, *233*, 65–73. [CrossRef]
29. Pan, C.-J.; Hou, Y.; Wang, Y.-N.; Gao, F.; Liu, T.; Hou, Y.-H.; Zhu, Y.-F.; Ye, W.; Wang, L.-R. Effects of Self-Assembly of 3-Phosphonopropionic Acid, 3-Aminopropyltrimethoxysilane and Dopamine on the Corrosion Behaviors and Biocompatibility of a Magnesium Alloy. *Mater. Sci. Eng. C* **2016**, *67*, 132–143. [CrossRef]
30. Hornberger, H.; Virtanen, S.; Boccaccini, A.R. Biomedical Coatings on Magnesium Alloys–A Review. *Acta Biomater.* **2012**, *8*, 2442–2455. [CrossRef]
31. Khan, S.A.; Miyashita, Y.; Mutoh, Y. Corrosion Fatigue Behavior of AM60 Magnesium Alloy with Anodizing Layer and Chemical-Conversion-Coating Layer: Corrosion Fatigue Behavior of Coated AM60 Magnesium Alloy. *Mater. Corros.* **2015**, *66*, 940–948. [CrossRef]
32. Liu, Z.; Gao, W. Electroless Nickel Plating on AZ91 Mg Alloy Substrate. *Surf. Coat. Technol.* **2006**, *200*, 5087–5093. [CrossRef]
33. Zhu, Y.; Wu, G.; Zhang, Y.-H.; Zhao, Q. Growth and Characterization of $Mg(OH)_2$ Film on Magnesium Alloy AZ31. *Appl. Surf. Sci.* **2011**, *257*, 6129–6137. [CrossRef]
34. Mao, L.; Shen, L.; Chen, J.; Wu, Y.; Kwak, M.; Lu, Y.; Xue, Q.; Pei, J.; Zhang, L.; Yuan, G.; et al. Enhanced Bioactivity of Mg–Nd–Zn–Zr Alloy Achieved with Nanoscale MgF_2 Surface for Vascular Stent Application. *ACS Appl. Mater. Interfaces* **2015**, *7*, 5320–5330. [CrossRef] [PubMed]
35. Liang, C.-S.; Lv, Z.-F.; Zhu, Y.-L.; Xu, S.-A.; Wang, H. Protection of Aluminium Foil AA8021 by Molybdate-Based Conversion Coatings. *Appl. Surf. Sci.* **2014**, *288*, 497–502. [CrossRef]
36. Mao, L.; Zhu, H.; Chen, L.; Zhou, H.; Yuan, G.; Song, C. Enhancement of Corrosion Resistance and Biocompatibility of Mg-Nd-Zn-Zr Alloy Achieved with Phosphate Coating for Vascular Stent Application. *J. Mater. Res. Technol.* **2020**, *9*, 6409–6419. [CrossRef]
37. Levy, G.; Aghion, E. Effect of Diffusion Coating of Nd on the Corrosion Resistance of Biodegradable Mg Implants in Simulated Physiological Electrolyte. *Acta Biomater.* **2013**, *9*, 8624–8630. [CrossRef]
38. Guo, L.; Zhang, F.; Song, L.; Zeng, R.-C.; Li, S.-Q.; Han, E.-H. Corrosion Resistance of Ceria/Polymethyltrimethoxysilane Modified Magnesium Hydroxide Coating on AZ31 Magnesium Alloy. *Surf. Coat. Technol.* **2017**, *328*, 121–133. [CrossRef]
39. Zheng, Y.F.; Gu, X.N.; Witte, F. Biodegradable Metals. *Mater. Sci. Eng. R: Rep.* **2014**, *77*, 1–34. [CrossRef]
40. Montemor, M.F.; Simões, A.M.; Ferreira, M.G.S.; Carmezim, M.J. Composition and Corrosion Resistance of Cerium Conversion Films on the AZ31 Magnesium Alloy and Its Relation to the Salt Anion. *Appl. Surf. Sci.* **2008**, *254*, 1806–1814. [CrossRef]
41. Bakhsheshi-Rad, H.R.; Hamzah, E.; Daroonparvar, M.; Saud, S.N.; Abdul-kadir, M.R. Bi-Layer Nano-TiO_2/FHA Composite Coatings on Mg–Zn–Ce Alloy Prepared by Combined Physical Vapour Deposition and Electrochemical Deposition Methods. *Vacuum* **2014**, *110*, 127–135. [CrossRef]
42. Chiu, K.Y.; Wong, M.H.; Cheng, F.T.; Man, H.C. Characterization and Corrosion Studies of Fluoride Conversion Coating on Degradable Mg Implants. *Surf. Coat. Technol.* **2007**, *202*, 590–598. [CrossRef]
43. Yao, H.B.; Li, Y.; Wee, A.T.S. Passivity Behavior of Melt-Spun Mg–Y Alloys. *Electrochim. Acta* **2003**, *48*, 4197–4204. [CrossRef]
44. Verdier, S.; Delalande, S.; van der Laak, N.; Metson, J.; Dalard, F. Monochromatized X-Ray Photoelectron Spectroscopy of the AM60 Magnesium Alloy Surface after Treatments in Fluoride-Based Ti and Zr Solutions. *Surf. Interface Anal.* **2005**, *37*, 509–516. [CrossRef]
45. Chen, J.; Song, Y.; Shan, D.; Han, E.-H. In Situ Growth of Mg–Al Hydrotalcite Conversion Film on AZ31 Magnesium Alloy. *Corros. Sci.* **2011**, *53*, 3281–3288. [CrossRef]
46. Zai, W.; Su, Y.; Man, H.C.; Lian, J.; Li, G. Effect of PH Value and Preparation Temperature on the Formation of Magnesium Phosphate Conversion Coatings on AZ31 Magnesium Alloy. *Appl. Surf. Sci.* **2019**, *492*, 314–327. [CrossRef]
47. Verdier, S.; van der Laak, N.; Delalande, S.; Metson, J.; Dalard, F. The Surface Reactivity of a Magnesium–Aluminium Alloy in Acidic Fluoride Solutions Studied by Electrochemical Techniques and XPS. *Appl. Surf. Sci.* **2004**, *235*, 513–524. [CrossRef]
48. Xu, L.; Pan, F.; Yu, G.; Yang, L.; Zhang, E.; Yang, K. In Vitro and in Vivo Evaluation of the Surface Bioactivity of a Calcium Phosphate Coated Magnesium Alloy. *Biomaterials* **2009**, *30*, 1512–1523. [CrossRef]
49. Zhang, S.; Li, Q.; Chen, B.; Yang, X. Preparation and Corrosion Resistance Studies of Nanometric Sol–Gel-Based CeO_2 Film with a Chromium-Free Pretreatment on AZ91D Magnesium Alloy. *Electrochim. Acta* **2010**, *55*, 870–877. [CrossRef]
50. Gray-Munro, J.E.; Strong, M. The Mechanism of Deposition of Calcium Phosphate Coatings from Solution onto Magnesium Alloy AZ31. *J. Biomed. Mater. Res.* **2009**, *90*, 339–350. [CrossRef]
51. Vandrovcová, M.; Bačáková, L. Adhesion, Growth and Differentiation of Osteoblasts on Surface-Modified Materials Developed for Bone Implants. *Physiol. Res.* **2011**, *60*, 403–417. [CrossRef]
52. Dou, Y.; Cai, S.; Ye, X.; Xu, G.; Huang, K.; Wang, X.; Ren, M. 45S5 Bioactive Glass–Ceramic Coated AZ31 Magnesium Alloy with Improved Corrosion Resistance. *Surf. Coat. Technol.* **2013**, *228*, 154–161. [CrossRef]
53. Yan, T.; Tan, L.; Xiong, D.; Liu, X.; Zhang, B.; Yang, K. Fluoride Treatment and In Vitro Corrosion Behavior of an AZ31B Magnesium Alloy. *Mater. Sci. Eng. C* **2010**, *30*, 740–748. [CrossRef]
54. Li, Q.; Zhu, P.; Chen, S.; Zhang, B.; Yang, K. *In Vitro* Study on Degradation of AZ31B Magnesium Alloy with Fluoride Conversion Coating. *Mater. Technol.* **2017**, *32*, 409–414. [CrossRef]
55. Song, G.; StJohn, D. Corrosion Behaviour of Magnesium in Ethylene Glycol. *Corros. Sci.* **2004**, *46*, 1381–1399. [CrossRef]

56. Barajas, J.D.; Joya, J.C.; Durán, K.S.; Hernández-Barrios, C.A.; Coy, A.E.; Viejo, F. Relationship between Microstructure and Formation-Biodegradation Mechanism of Fluoride Conversion Coatings Synthesised on the AZ31 Magnesium Alloy. *Surf. Coat. Technol.* **2019**, *374*, 424–436. [CrossRef]
57. Chen, Y.; Xu, Z.; Smith, C.; Sankar, J. Recent Advances on the Development of Magnesium Alloys for Biodegradable Implants. *Acta Biomater.* **2014**, *10*, 4561–4573. [CrossRef] [PubMed]
58. Brash, J.L. Exploiting the Current Paradigm of Blood–Material Interactions for the Rational Design of Blood-Compatible Materials. *J. Biomater. Sci. Polym. Ed.* **2000**, *11*, 1135–1146. [CrossRef]
59. Nakanishi, K.; Sakiyama, T.; Imamura, K. On the Adsorption of Proteins on Solid Surfaces, a Common but Very Complicated Phenomenon. *Bioengineering* **2001**, *91*, 233–244. [CrossRef]
60. Liu, C.-L.; Zhang, Y.; Zhang, C.-Y.; Wang, W.; Huang, W.-J.; Chu, P.K. Synergistic Effect of Chloride Ion and Albumin on the Corrosion of Pure Magnesium. *Front. Mater. Sci.* **2014**, *8*, 244–255. [CrossRef]
61. Roach, P.; Farrar, D.; Perry, C.C. Interpretation of Protein Adsorption: Surface-Induced Conformational Changes. *J. Am. Chem. Soc.* **2005**, *127*, 8168–8173. [CrossRef]
62. Zhen, Z.; Liu, X.; Huang, T.; Xi, T.; Zheng, Y. Hemolysis and Cytotoxicity Mechanisms of Biodegradable Magnesium and Its Alloys. *Mater. Sci. Eng. C* **2015**, *46*, 202–206. [CrossRef] [PubMed]
63. Zheng, Q.; Sun, Z.; Wang, Z.; Duan, T.; Xu, K.; Cai, M.; Wang, B. Corrosion and Biocompatibility Behaviours of Microarc Oxidation/Phytic Acid Coated Magnesium Alloy Clips for Use in Cholecystectomy in a Rabbit Model. *RSC Adv.* **2021**, *11*, 20730–20736. [CrossRef] [PubMed]
64. Gu, X.; Zheng, Y.; Cheng, Y.; Zhong, S.; Xi, T. In Vitro Corrosion and Biocompatibility of Binary Magnesium Alloys. *Biomaterials* **2009**, *30*, 484–498. [CrossRef] [PubMed]
65. Gorbet, M.B.; Sefton, M.V. Biomaterial-Associated Thrombosis: Roles of Coagulation Factors, Complement, Platelets and Leukocytes. *Biomaterials* **2004**, *25*, 5681–5703. [CrossRef] [PubMed]
66. Kim, B.-R.; Nguyen, T.B.L.; Min, Y.-K.; Lee, B.-T. *In Vitro* and *In Vivo* Studies of BMP-2-Loaded PCL–Gelatin–BCP Electrospun Scaffolds. *Tissue Eng. Part A* **2014**, *20*, 3279–3289. [CrossRef]
67. Zhang, L.-C.; Xu, M.; Hu, Y.-D.; Gao, F.; Gong, T.; Liu, T.; Li, X.; Pan, C.-J. Biofunctionization of Biodegradable Magnesium Alloy to Improve the In Vitro Corrosion Resistance and Biocompatibility. *Appl. Surf. Sci.* **2018**, *451*, 20–31. [CrossRef]
68. Trinidad, J.; Arruebarrena, G.; Marco, I.; Hurtado, I.; Sáenz de Argandoña, E. Effectivity of Fluoride Treatment on Hydrogen and Corrosion Product Generation in Temporal Implants for Different Magnesium Alloys. *Proc. Inst. Mech. Eng. H* **2013**, *227*, 1301–1311. [CrossRef]
69. Vermette, P. Effects of Surface Properties and Bioactivation of Biomaterials on Endothelial Cells. *Front. Biosci.* **2010**, *S2*, 239–255. [CrossRef]
70. Yu, W.; Zhao, H.; Ding, Z.; Zhang, Z.; Sun, B.; Shen, J.; Chen, S.; Zhang, B.; Yang, K.; Liu, M.; et al. In Vitro and In Vivo Evaluation of MgF_2 Coated AZ31 Magnesium Alloy Porous Scaffolds for Bone Regeneration. *Colloids Surf. B Biointerfaces* **2017**, *149*, 330–340. [CrossRef]
71. Horbett, T.A.; Waldburger, J.J.; Ratner, B.D.; Hoffman, A.S. Cell Adhesion to a Series of Hydrophili-Hydrophobic Copolymers Studies with a Spinning Disc Apparatus. *J. Biomed. Mater. Res.* **1988**, *22*, 383–404. [CrossRef]
72. Pereda, M.D.; Alonso, C.; Gamero, M.; del Valle, J.A.; Fernández Lorenzo de Mele, M. Comparative Study of Fluoride Conversion Coatings Formed on Biodegradable Powder Metallurgy Mg: The Effect of Chlorides at Physiological Level. *Mater. Sci. Eng. C* **2011**, *31*, 858–865. [CrossRef]

metals

Article

Composite Coating Prepared with Ferulic Acid to Improve the Corrosion Resistance and Blood Compatibility of Magnesium Alloy

Zhijin Han [1], Haojie Guo [1], Yifan Zhou [1,*], Liguo Wang [1], Kun Zhang [2] and Jing-an Li [1,*]

1 School of Materials Science and Engineering, Henan Key Laboratory of Advanced Magnesium Alloys, Zhengzhou University, 100 Science Road, Zhengzhou 450001, China; hzj201912192012600@gs.zzu.edu.cn (Z.H.); 20212192013527@gs.zzu.edu.cn (H.G.); lgwang@zzu.edu.cn (L.W.)

2 School of Life Science, Zhengzhou University, 100 Science Road, Zhengzhou 450001, China; zhangkun@zzu.edu.cn

* Correspondence: zhouyf@zzu.edu.cn (Y.Z.); lijingan@zzu.edu.cn (J.-a.L.)

Abstract: Magnesium (Mg) alloy has been used for medical vascular stents because of its good biocompatibility and degradability, but its rapid degradation and poor blood compatibility limits its further application. In this study, ferulic acid (FA) was conjugated onto the polydopamine (PDA) deposited Mg-Zn-Y-Nd alloy to prepare a PDA/FA multi-functional coating with better corrosion resistance and blood compatibility. The results suggest that the PDA/FA coating possessed potential application for surface modification of a medical Mg alloy.

Keywords: cardiovascular stent; Mg-Zn-Y-Nd alloy; blood compatibility; corrosion resistance; ferulic acid

check for updates

Citation: Han, Z.; Guo, H.; Zhou, Y.; Wang, L.; Zhang, K.; Li, J.-a. Composite Coating Prepared with Ferulic Acid to Improve the Corrosion Resistance and Blood Compatibility of Magnesium Alloy. *Metals* **2022**, *12*, 545. https://doi.org/10.3390/met12040545

Academic Editor: Sebastian Feliu Batlle

Received: 23 February 2022
Accepted: 21 March 2022
Published: 23 March 2022

1. Introduction

Cardiovascular disease (CVD) remains a major cause of health loss for all regions of the world [1]. The most common treatment strategy for cardiovascular disease has been implantation of vascular stents [2]. Vascular stents are divided into drug-eluting stents (DES), bare metal stents (BMS), and degradable stents. Traditional metal vascular stents (such as cobalt chromium alloy and 316 L stainless steel) are undegradable and face the risk of secondary surgery. Drug-eluting stents are superior to conventional non-degradable scaffolds in reducing major adverse clinical events [3]. Although DES can significantly reduce the incidence of restenosis, it still has some disadvantages, such as late-stent thrombosis (LST), accelerated atherosclerosis, and delayed endothelialization. Anti-proliferative drugs (sirolimus, paclitaxel, etc.) inhibit endothelial cell proliferation [4–6]. The degradation of degradable polymer stents will produce local acidity, resulting in severe inflammation and intimal hyperplasia [7]. Therefore, the search for biodegradable metal stents has become a hot point in the field of vascular disease research [8].

Magnesium (Mg) alloy cardiovascular stents have attracted much attention because of their excellent properties, such as excellent biocompatibility and degradability [9–11]. However, the application of the Mg alloy in cardiovascular stents still faces some problems, such as rapid degradation rate, thrombosis, and insufficient endothelization. Consequently, it is necessary to reduce the corrosion rate, blood compatibility, and cell compatibility of Mg alloys by surface modification and the introduction of loaded drugs to prepare functional coatings. At present, surface-coating technology has been widely used in magnesium alloy, because it effectively decreases the degradation rate of the Mg alloy substrate and improves its biocompatibility. In addition, loading drugs into the coating can also obtain specific functions. Ferulic acid (FA) is the main effective component of Angelica sinensis, ferula, and other traditional Chinese medicines. It exists in fruits and vegetables such as corn and rice bran [12]. It has the physiological characteristics of being anti-inflammatory, antibacterial, anti-virus, anti-aging, and anti-cancer [13,14]. In addition, it also has the function of anti-platelet aggregation. Its sodium salt has been used in the adjuvant treatment of coronary heart disease, atherosclerosis, and other diseases [15]. More importantly, studies have shown that FA can promote angiogenesis [16] and endothelial cell proliferation [17], and it can inhibit the proliferation of smooth muscle cells [18]. Zhang et al. found that FA can improve the blood compatibility of poly (3-hydroxybutyrateco-3-hydroxyhexanoate) (PHBHHx) membrane [19]. These characteristics show that FA is an ideal drug for cardiovascular stents. Polydopamine (PDA) has the characteristics of easy adhesion, easy preparation, and good biocompatibility, and it has a functional group amino, which can play the role of double-sided adhesive. However, it is difficult to be applied in practice due to poor blood compatibility. PDA is rich in functional groups such as amino groups [20–22]. Therefore, PDA can be selected to load FA [23]. The preparation of functional coatings by surface treatment is a common method by which to enhance the comprehensive properties of Mg alloys, such as anodic oxidation [8,24], fluorination [25], phosphate treatment [26], electro-grafting [27,28], silylation [29–31], and alkali heat treatment [19,20,32]. Among them, alkali heat treatment has the characteristics of having a simple method, being easy to prepare, having a low cost, and a degradable coating, which is convenient for the next coating preparation and conducive to cell adhesion [19]. To solve the lack of corrosion resistance of the Mg alloy, an alkaline thermal coating could be prepared under the PDA coating.

In this work, Mg-OH/PDA/FA composite coatings were successfully prepared on the Mg-Zn-Y-Nd alloy. The Mg-Zn-Y-Nd alloy was treated with alkali heat treatment, then modified with dopamine, and the PDA coating was obtained as the fixed layer. FA were grafted on PDA through carbodiimide chemistry and the Mannich reaction. It has been reported that FA is dose-dependent, i.e., different concentrations of phenolic hydroxyl and polyphenol hydroxyl will have different effects on the growth behavior of cells [33]. The PDA coating was treated with FA for three different time periods to conjugate different doses of FA.

2. Materials and Methods

2.1. Chemicals and Materials

The Mg-Zn-Y-Nd alloy (Mg-2.0Zn-0.46Y-0.5Nd, extruded) used as the basal material was obtained from Henan Key Laboratory of Advanced Magnesium Alloy (Zhengzhou, China). Dopamine hydrochloride and Trans-Ferulic acid were purchased from Aladdin Co., (Shanghai, China). N-(3-(dimethylamino) propyl)-N0-ethylcarbodiimide hydro-chloride (EDC), n-hydroxysuccinimide (NHS), and 2—(N-morpholine) ethanesulfonic acid (MES) were purchased from Shanghai Macklin Biochemical Co., Ltd. (Shanghai, China).

2.2. Preparation of PDA-FA Composite Coating on Mg-Zn-Y-Nd Alloy

The Mg-Zn-Y-Nd alloy substrates were polished and cleaned as reported previously [34]. Then, the Mg-Zn-Y-Nd alloy substrates were promptly soaked in 3 M NaOH solution for 8 h at 60 °C. The pretreated substrates were rinsed with deionized water (dH_2O) 3 times. The samples, after alkali treatment, were coded as Mg-OH. The treated substrates were then soaked in 2 mg/mL dopamine Tris buffer solution (pH 8.5) at 37 °C for 24 h to obtain a PDA coating, The PDA coating was rinsed 3 times with dH_2O and dried with hot air, and the samples were coded as PDA. Afterward, FA was immersed in a water-soluble carbodiimide (WSC) solution (pH 5.4) containing 1 mg/mL N-(3-dimethylaminopropyl)-N′-ethylcarbodiimide (EDC), 0.24 mg/mL nhydroxysuccinimide (NHS), and 2 mg/mL 2-(N-morpholino) ethanesulfonic acid hydrate (MES) for 30 min at 37 °C in order to activate the carboxyl group of ferulic acid. Then, the PDA samples were soaked in WSC solution at 37 °C for three different time gradients. The three times gradients of the activated FA carboxyl group reacting with the PDA amino group were coded as FA-1 h, FA-3 h and FA-6 h. The preparation process of Mg-OH/PDA/FA coating was shown in Figure 1.

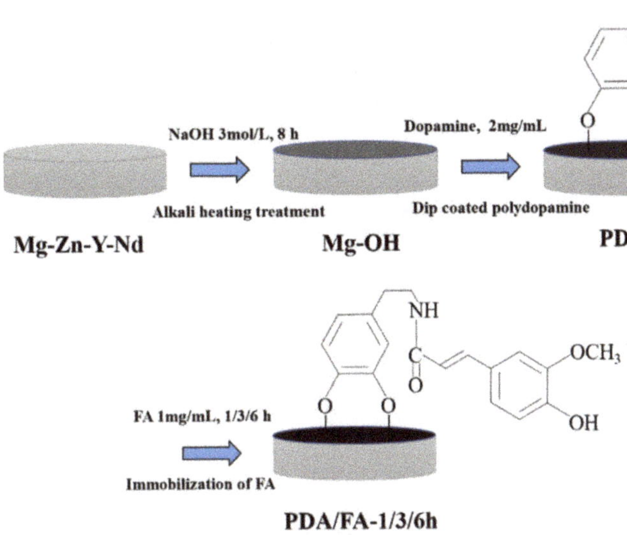

Figure 1. The scheme for preparing Mg-OH/PDA/FA composite coating.

2.3. *Characterization of the Mg-OH/PDA/FA Composite Coating*

The surface morphology of the Mg-Zn-Y-Nd, Mg-OH, PDA, FA-1 h, FA-3 h, FA-6 h samples was observed by scanning electron microscope (SEM, FEI Quanta200, Eindhoven, The Netherlands) [22]. The composition and functional groups of Mg-Zn-Y-Nd, Mg-OH, PDA, FA-1 h, FA-3 h, FA-6 h samples were analyzed by total reflection Fourier transform infrared spectroscopy (FTIR, Bruker VERTEX 70v, Karlsruhe, Germany) [35]. The water contact angle of Mg-Zn-Y-Nd, Mg-OH, PDA, FA-1 h, FA-3 h, FA-6 h samples was measured by water contact angle-measuring equipment (DSA 100, Krüss, GmbH, Hamburg, Germany) at room temperature, and then the average of six readings was calculated from each sample by DSA 1.8 software (Version 1.8, Krüss, GmbH, Hamburg, Germany) [36]. The micromorphology and roughness of Mg-Zn-Y-Nd, Mg-OH, PDA, FA-1 h, FA-3 h, and FA-6 h were measured by atomic force microscopy (AFM, Bruker/mnltimode, Billerica, MA, USA) [37]. The chemical composition was further studied by X-ray photoelectron spectroscopy (XPS, Axis Supra, Kyoto, Japan) [38]. Electrochemical corrosion tests including potential dynamic polarization (potentiodynamic polarization plots, scanning rate: 0.0005 v/s, 20 min) and electrochemical impedance spectroscopy (Nyquist EIS spectra, Zhengzhou Shi Ruisi Instrument Technology Co., Ltd., Zhengzhou, China) were carried out by RST5200f electrochemical workstation (Shiruisi, Zhengzhou, China) with a three-electrode system in simulated body fluid (SBF) solution [39].

2.4. *Blood Compatibility Tests*

2.4.1. Adhesion and Degeneration of Fibrinogen

The fibrinogen denaturation experiment was carried out by quantification of γ-chain exposure by using the ELISA method [40]. In short, fresh blood with heparin sodium as a anticoagulant was centrifuged at $3000\times g$ rpm to acquire poor-platelet plasma (PPP). Then the sample to be tested was placed on a 24-well plate, and 50 μL PPP was added to each sample, after which it stood at 37 °C for 1 h. After washing with normal saline 3 times, HRP-labeled goat anti-human fibrinogen-binding antibody was added to each sample. Finally, the absorbance at 450 nm was measured by an enzyme-labeling instrument.

2.4.2. Platelet Adhesion Test

Fresh whole blood was taken from healthy volunteers. Sodium citrate was used as an anticoagulant, centrifuged at 1500× g r/min for 10 min, and the supernatant was absorbed to obtain platelet-rich plasma (PRP). After that, the samples were soaked in 50 μL PRP and incubated for 1 h at 37 °C. Finally, all samples were washed 3 times with normal saline for 3 min each time to clear non-adherent platelets, then fixed with 4% paraformaldehyde, and finally dehydrated and dried with 25, 50, 75, 90, and 100% ethanol solutions. The surface morphology was observed by SEM [41].

2.4.3. Hemolysis Test

Fresh blood and normal saline were mixed in the ratio of 4:5 to prepare diluted blood, and then the sample was immersed to be tested for 30 min in 37 °C normal saline. Then the diluted blood was added to 37 °C normal saline and ultrapure water for 1 h as a negative control group and a positive control group. Finally, the solution was sucked out and centrifuged at 2500× g rpm for 5 min. Finally, the absorbance at 545 nm wavelength was measured by an enzyme-labeling instrument. The hemolysis ratio (HR) is calculated as follows:

$$HR = \frac{OD_t - OD_n}{OD_p - OD_n} \times 100\%, \tag{1}$$

where OD_t represents the optical density value of the sample group, OD_n represents the optical density value of the negative control group, and OD_p represents the optical density value of the positive group. If the hemolysis rate exceeds 5%, it means that hemolysis reaction may occur [42]. Otherwise, it indicates that the material meets the requirements of medical materials.

2.5. DPPH Free Radical Content Test

The scavenging ability of free radicals of each sample was determined by the 2,2-diphenyl-1-pyridyl (DPPH) method and the DPPH free radical decolorization method from purple to yellow. That is, the sample was immersed in 2 mL DPPH ethanol solution (0.02 mmol/mL) and then treated in the dark at 37 °C for 1 h. Then, the absorbance at 517 nm was measured by enzyme plate absorbance meter (synergyh1, BioTek, Vermont, VT, USA).

2.6. Statistical Analysis

The results were indicated as mean ± standard deviation, and the difference was statistically significant if the $p < 0.05$.

3. Results and Discussion

3.1. Surface Characterization

Figure 2 shows the SEM morphology and surface element composition of each sample. After polishing, the surface of the Mg alloy is smooth. A dense and uniform $Mg(OH)_2$ passivation layer is formed after alkali heat treatment, and the O element appears in the coating. After depositing a PDA film, C and N elements appear in the coating, and regular cracks appear on the surface, which may be caused by the cracks of the alkaline thermal coating itself or the self-polymerization of dopamine. When FA was fixed for 1 h, some cracks were filled. At 3 h, the cracks were basically filled up, and some particles are generated. At 6 h, more detailed and irregular cracks were produced, which may be caused by the overreaction between FA and PDA. With the increase of fixed time, the content of the C element gradually increased and the content of the O element decreased.

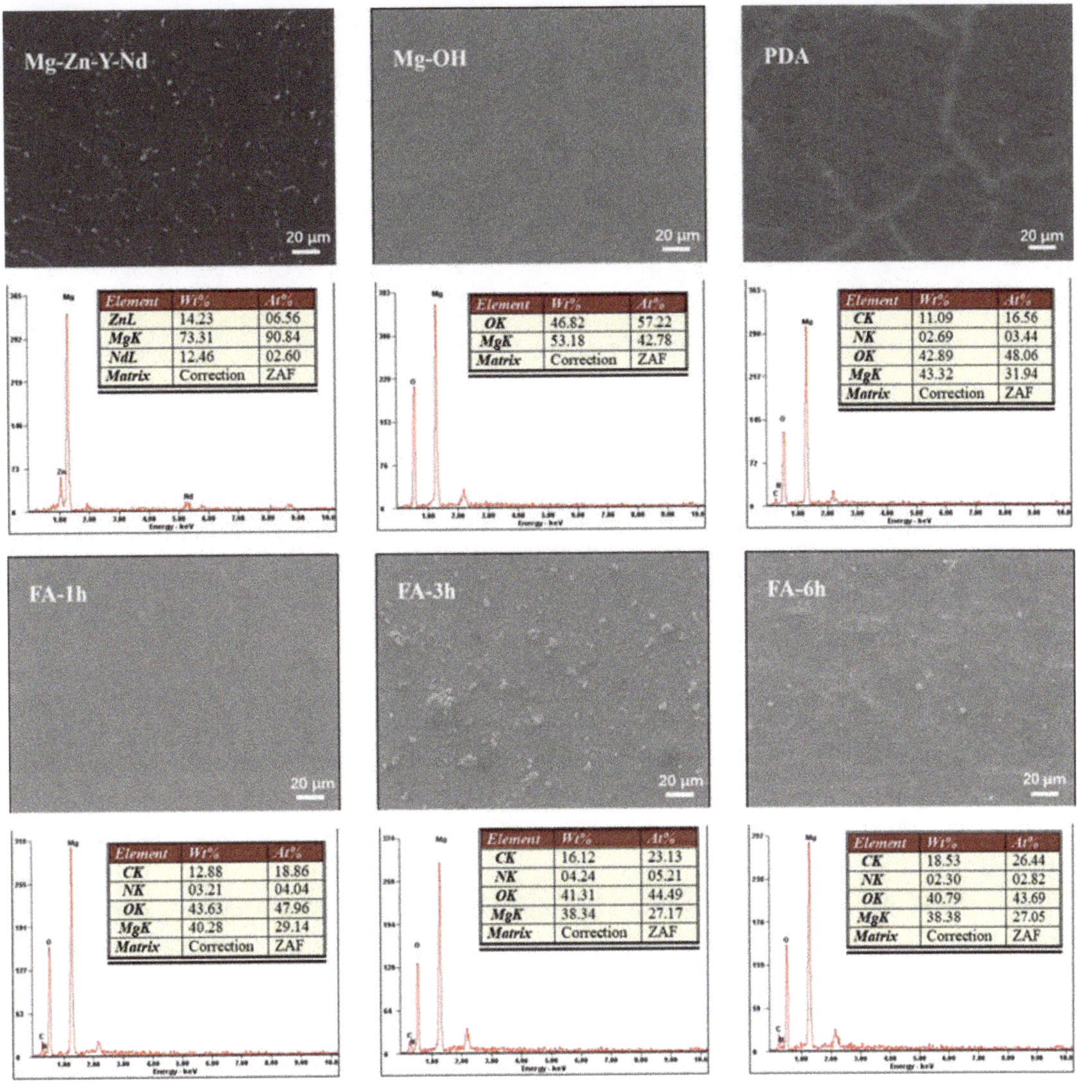

Figure 2. Surface SEM morphology and EDS spectrum of samples.

Here, the chemical composition and structure of PDA before and after FA fixation were analyzed by infrared technology and XPS technology, and the successful fixation of FA was verified. Figure 3 shows the infrared spectrum of the samples. Alkali heat treatment can introduce a large amount of—OH on the surface, so it has obvious infrared absorption of Mg-OH at 3704 cm^{-1}. After dipping in dopamine, the substrate presented several peaks at 1204.1099 cm^{-1}, which are characteristic peaks for stretching and bending vibrations of C-O, C-N of the PDA molecules on the material surface, respectively. Further modifications, in case of FA grafting, include the absorption band around 1300–800 cm^{-1} being reinforced. Three new peaks at 1671 cm^{-1}, 1530 cm^{-1}, and 1365 cm^{-1} were observed which are characteristic peaks for C=O stretching, N-H bending, and C-N stretching of the amide bond. This preliminarily confirmed the successful fixation of FA.

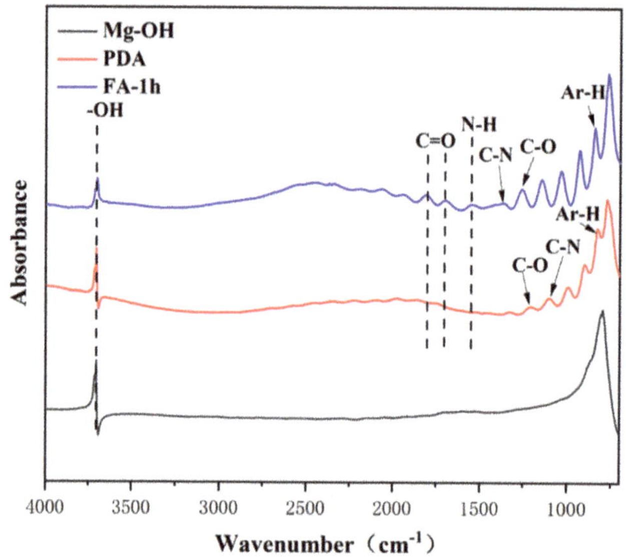

Figure 3. Infrared spectra of different samples.

As shown in Figure 4, the fixed FA will significantly reduce the N peak (399.5 eV) and O peak (531.5 eV) and increase the C peak (248.8 eV). In the C1s spectrum, 286 and 288.4 eV peaks were enhanced and ascribed to the increase of phenolic hydroxyl groups immobilized by FA and the formation of amide bonds. New peaks appeared at 400.4 and 531.5 eV in N1s and O1s, and the peaks moved toward high-binding energy, which is ascribed to the formation of amide bond-N-C=O.

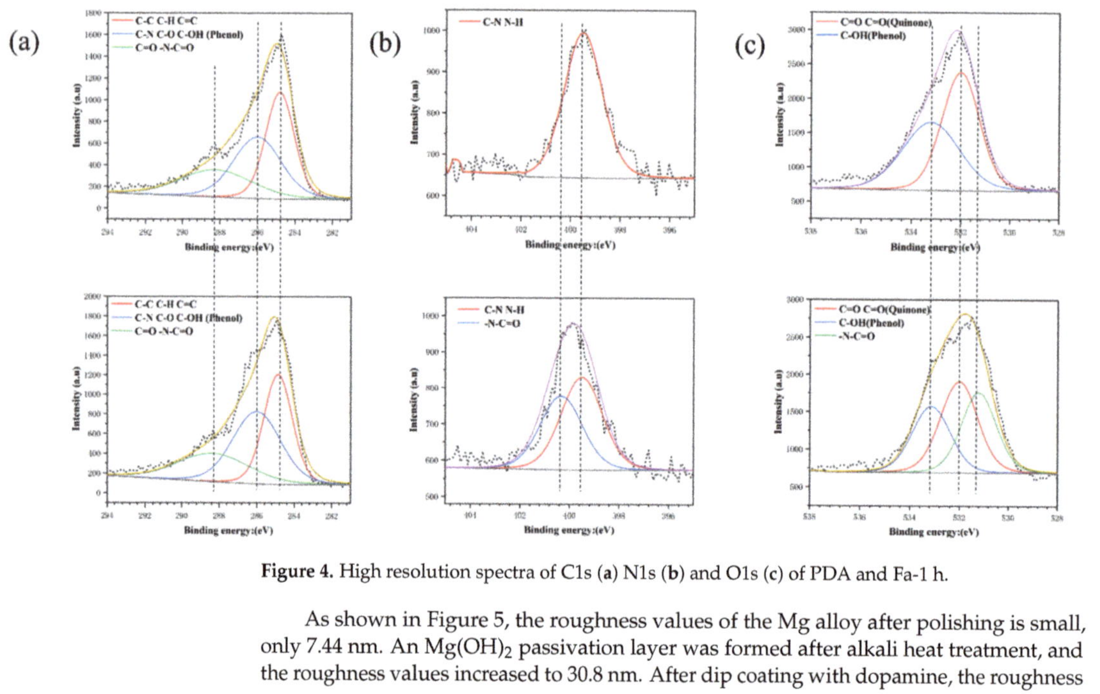

Figure 4. High resolution spectra of C1s (**a**) N1s (**b**) and O1s (**c**) of PDA and Fa-1 h.

As shown in Figure 5, the roughness values of the Mg alloy after polishing is small, only 7.44 nm. An $Mg(OH)_2$ passivation layer was formed after alkali heat treatment, and the roughness values increased to 30.8 nm. After dip coating with dopamine, the roughness values continued to increase to 33.4 nm, which may be related to cracks and particles in the dopamine coating. However, after fixing FA, the roughness values increases to 35.9 nm, and with the increase of fixing time, the roughness values gradually decrease to 29.5 nm at 3 h and 26.7 nm at 6 h, which may be ascribed to FA filling the cracks of PDA coating.

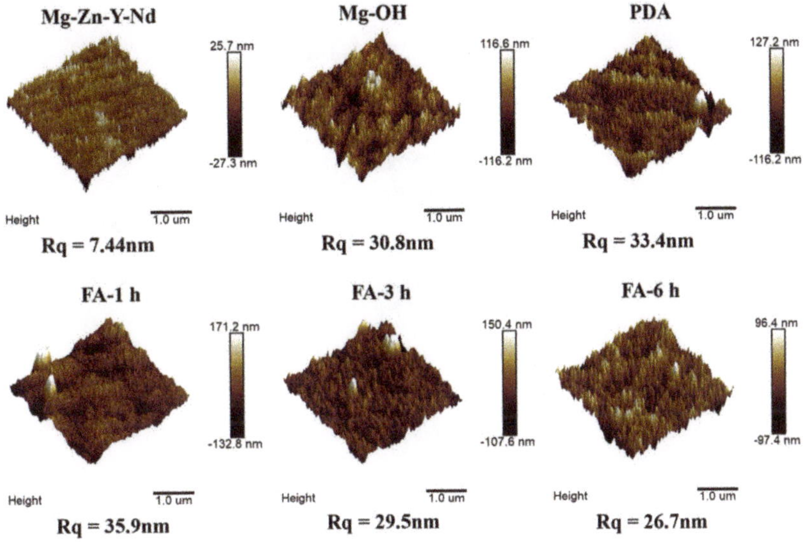

Figure 5. Atomic force topography and roughness of samples.

As shown in Figure 6, the Mg alloy had good hydrophilicity, and the water contact angle was only 28.3°, which may be due to low roughness and overpolishing. After alkali heat treatment, the water contact angle increased to 45.7°, which may be due to the formation of the $Mg(OH)_2$ passivation layer and the increase of roughness. After dip coating with dopamine, due to the addition of hydrophilic groups such as dopamine amino group and phenolic hydroxyl group, the water contact angle decreased significantly to 12°. With the increase of FA fixation time, the water contact angle gradually increased, which may be due to the addition of C=C hydrophobic groups. Although amide bonds were formed, the hydrophilicity was still slightly reduced, to 18.6°, 24.5°, and 37° respectively, but they were far less than 90°, showing hydrophilicity.

3.2. Corrosion Resistance Tests

As shown in Figure 7 and Table 1, the corrosion resistance of the Mg alloy was enhanced after the PDA/FA coating was prepared. The self-corrosion potential first increased and then decreased. After alkali heat treatment, the self-corrosion current decreased by two orders of magnitude compared with the Mg alloy. After dip-coating with dopamine, the self-corrosion current decreased by an order of magnitude compared with alkali heat treatment. After FA was fixed, the self-corrosion current first decreased and then increased with the increase of time. The variation law of the self-corrosion current was Fa-3 h > Fa-1 h > PDA > Mg-OH > Fa-6 h > Mg-Zn-Y-Nd. The variation law of the self-corrosion potential was Fa-3 h > Fa-1 h > PDA > Fa-6 h > Mg-OH > Mg-Zn-Y-Nd.

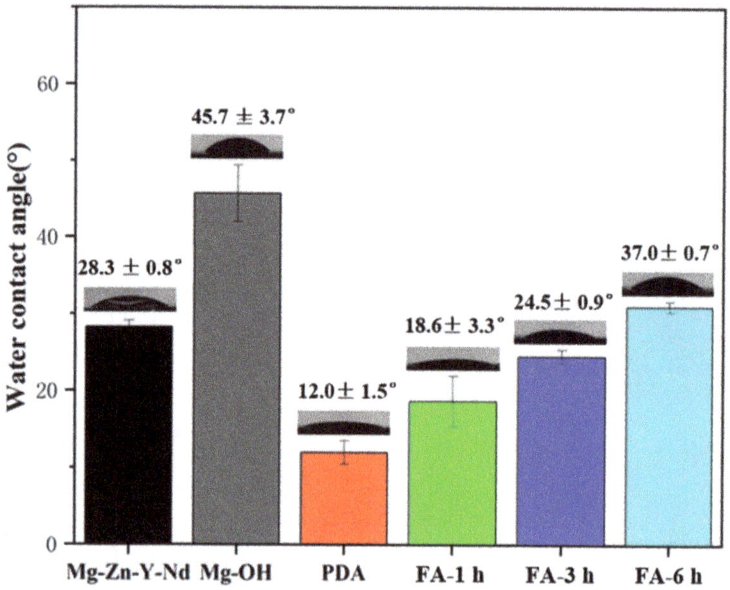

Figure 6. Water contact angle of samples (mean ± SD, n = 6).

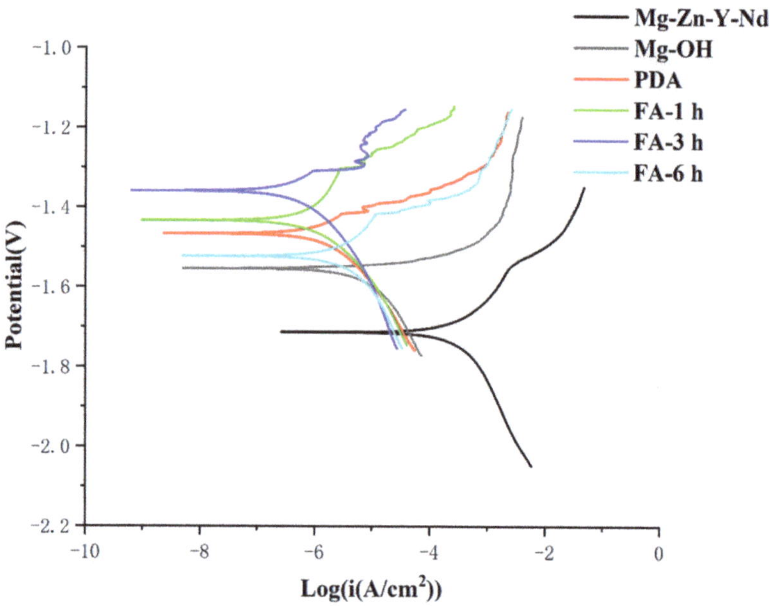

Figure 7. Potentiodynamic polarization plots of samples.

Table 1. Corrosion potential (Ecorr) and current density (Icorr) of samples.

Samples	E_{corr} (V)	I_{corr} (A/cm^2)
Mg-Zn-Y-Nd	−1.71	1.08×10^{-4}
Mg-OH	−1.55	2.47×10^{-6}
PDA	−1.47	6.50×10^{-7}
FA-1 h	−1.43	3.44×10^{-7}
FA-3 h	−1.36	2.06×10^{-7}
FA-6 h	−1.52	1.24×10^{-6}

In the low-frequency region, the intersection of the real part of the impedance and the x-axis reflected the impedance value of the sample. As shown in Figure 8, it can be found from the figure that the impedance value of Mg-Zn-Y-Nd was low, which was 129 Ohm, while the impedance values of Mg-OH and PDA samples were close, which were 5580 and 7080 Ohm, respectively. Moreover, the impedance of Fa-1 h, Fa-3 h, and Fa-6 h were 10,300, 10,900, and 12,200 Ohm, which was greater than that of the PDA coating and alkaline thermal coating, indicating that PDA/FA coating increased the corrosion resistance of the Mg alloy. The order of impedance was Fa-6 h > Fa-3 h > Fa-1 h > PDA > Mg-OH > Mg-Zn-Y-Nd. It is worth noting that the Fa-6 h impedance radius was larger, indicating good corrosion resistance, which may be due to slow corrosion in steady state.

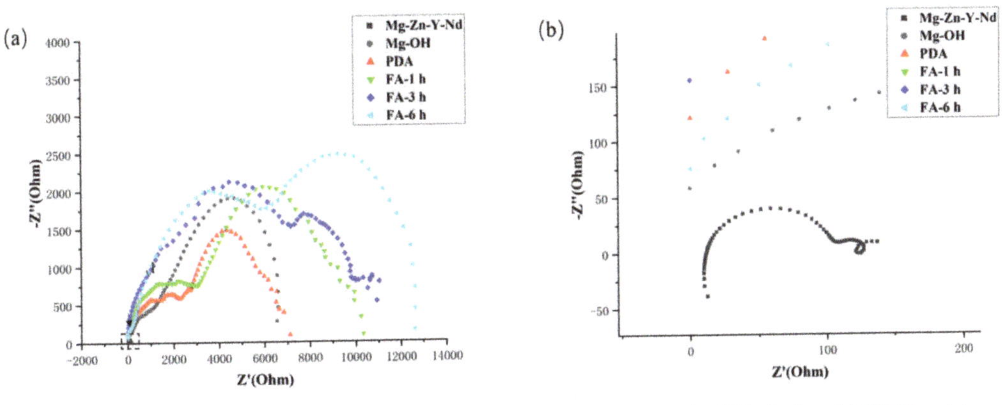

Figure 8. (**a**) EIS Nyquist spectra of the samples in the SBF solution; (**b**) EIS Nyquist spectra of Mg-Zn-Y-Nd in SBF solution.

3.3. Blood Compatibility Tests

As shown in Figure 9, the hemolysis rate of each sample was less than 5%, indicating that no hemolysis occurred. The order of hemolysis rate from largest to smallest was Mg-OH > Mg-Zn-Y-Nd > Fa-6 h > PDA > Fa-1 h > Fa-3 h. After FA was fixed, the hemolysis rate of the sample decreased. Although hydrothermal treatment improved the corrosion resistance of the Mg-Zn-Y-Nd alloy (Figures 7 and 8, Table 1), its hemolysis rate also further increased. FA coating can obviously significantly reduce the hemolysis rate of Mg-OH samples, but with the extension of coating preparation time, the hemolysis rate of samples may have the risk of rebound.

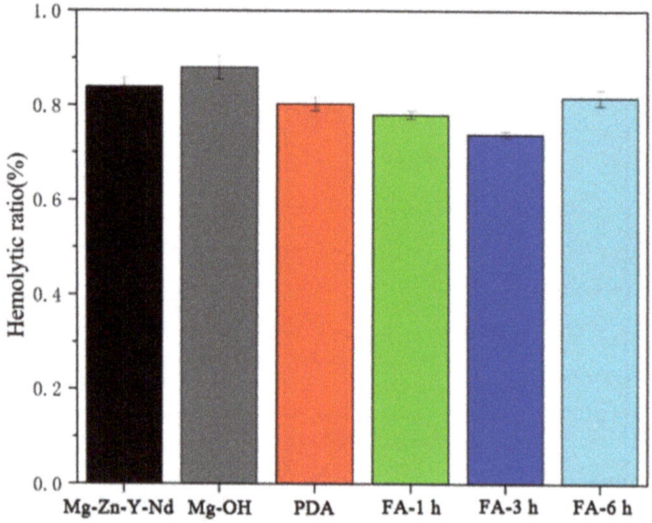

Figure 9. Hemolysis rate of each sample (mean ± SD, n = 3).

As shown in Figure 10, the numbers of adhesion and denaturation of fibrinogen (FGN) on the Mg alloy and alkali heat sample were larger, and the numbers of adhesion and denaturation of FGN after fixing FA were greatly reduced. The order of FGN adhesion from largest to smallest was Mg-OH > Mg-Zn-Y-Nd > Fa-1 h > Fa-3 h > Fa-6 h > PDA. The order of FGN denaturation quantity from largest to smallest is Mg-Zn-Y-Nd > Mg-OH > Fa-6 h > PDA > Fa-3 h > Fa-1 h. This indicated that PDA/FA coating can inhibit the adhesion and denaturation of FGN.

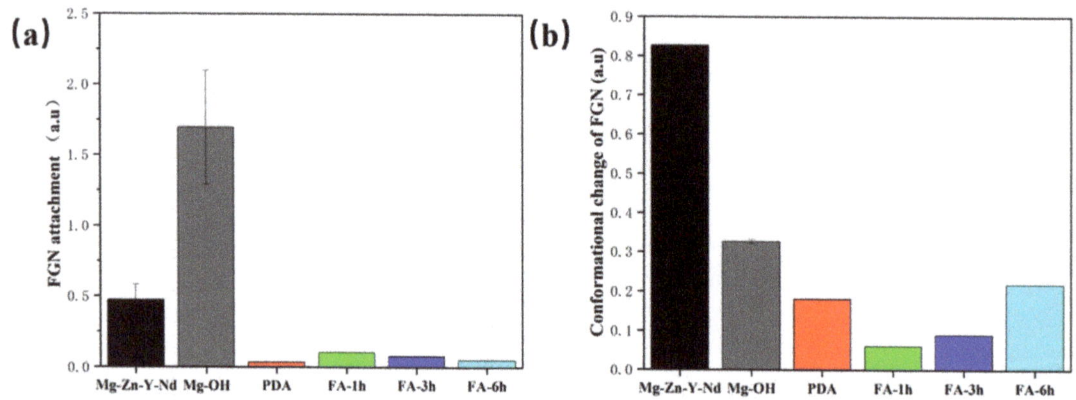

Figure 10. Adhesion (**a**) and denaturation (**b**) of FGN in each sample (mean ± SD, n = 3).

Research shows that Mg alloys have poor blood compatibility due to rapid degradation [43]. As shown in Figure 11, consistent with the literature, the Mg alloy was corroded, and a large number of platelets gathered on the surface, indicating that the Mg alloy may induce platelet activation. On the PDA coating, the platelets were partially aggregated, interconnected, and they extended pseudopodia showing a flat shape and a fully activated state, which indicated that the anti-platelet aggregation ability of PDA coating was poor. This may be because the PDA coating has good hydrophilicity and contains a large number of positive charges, which will adsorb negatively charged platelets and fibrin. After FA fixation, the number of platelets decreased significantly at 1 h, and some showed a spiny structure and tended to be activated. At 3 h, platelets were not spherical and showed an inactive state. At 6 h, the whole platelet still showed a spherical inactive state, but some platelets protrude pseudopodia. This indicates that FA may have good and obvious anti-platelet adhesion ability by inhibiting FGN adhesion and degeneration, or that FA inhibits platelet activity by antagonizing thromboxane.

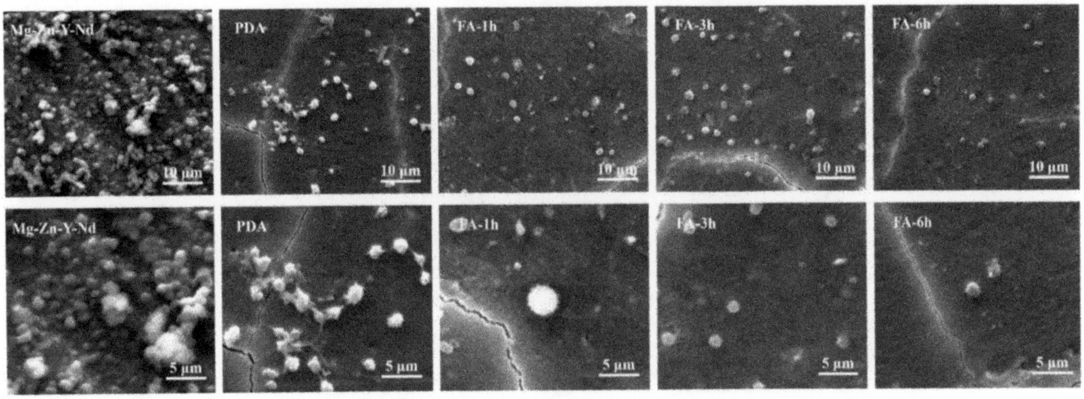

Figure 11. Platelet adhesion morphology of each sample.

The grafting time of FA corresponds to its amount on the surface. Generally, more molecules will be fixed on the surface after a longer grafting time, but this does not mean that a coating composed of more molecules will possess better functions [34]. In this study, the FA-6 h samples showed fewer platelets compared with FA-1 h and FA-3 h, but higher FGN denaturation compared to FA-1 h and FA-3 h. The platelet activation of FA-6 h was also more severe than FA-3 h. The reason may be due to its excessive preparation time. The weaker corrosion resistance of FA-6 h compared to that of FA-1 h and FA-3 h in Figure 7 and in Table 1 also indicated that its degradation behavior also affected blood compatibility.

3.4. DPPH Free Radical Content Test

The DPPH method was used to evaluate its free radical scavenging ability (Figure 12). The content of free radicals in the blank control group was the highest, and the content of free radicals decreased after adding the sample. The order of free radical scavenging ability was Fa-1 h > Fa-3 h > Fa-6 h > Mg-Zn-Y-Nd > PDA. The DPPH experiment showed that compared with the Mg alloy and PDA coating, FA fixation could significantly improve the free radical scavenging ability of the sample. The data of scavenging free radicals show that FA fixation can significantly improve the antioxidant capacity of the Mg alloy and PDA coating, which means that the coating can provide a good environment for the adhesion and proliferation of endothelial cells.

An ideal Mg alloy cardiovascular stent needs to reduce the adsorption of fibrin, promote the adhesion and proliferation of endothelial cells, inhibit the adhesion and proliferation of smooth muscle cells, and provide a good microenvironment for the implantation of cardiovascular stents. Therefore, surface modification and loading of new drugs is very important, and ferulic acid has the above functions. In this work, we used ferulic acid, a cardiovascular drug and antioxidant, as well as an extract of traditional Chinese medicine, combined with polydopamine to obtain a uniform and dense coating. After preparation, the coating has good oxidation resistance, which provides a new solution for the surface modification of Mg alloy cardiovascular stents.

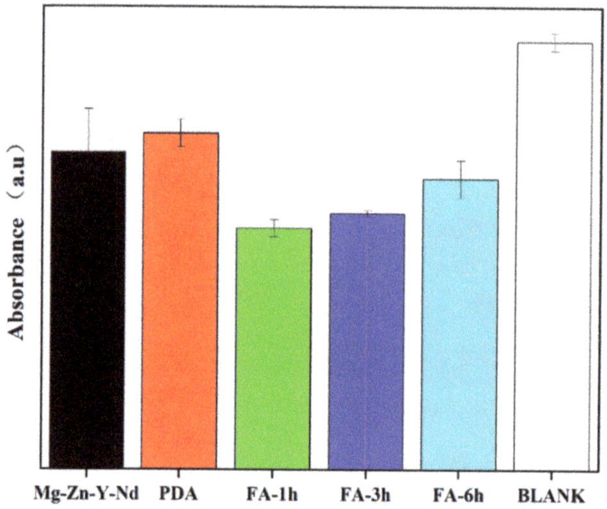

Figure 12. DPPH radical content of samples.

4. Conclusions

In this paper, the surface of Mg-Zn-Y-Nd alloy was modified by preparing a PDA/FA functional coating, which improved the corrosion resistance and biocompatibility of an Mg alloy. The modification of uniform PDA/FA coating makes the Mg alloy have better corrosion resistance, improves the anticoagulant ability, and widens the application of Mg alloy. It is worth noting that Fa-3 h has good corrosion resistance and blood compatibility. In summary, a functional composite coating is studied, which provides a new idea for the application of the medical Mg alloy.

Author Contributions: Conceptualization, Z.H. and Y.Z.; methodology, Z.H. and H.G.; software, Z.H. and H.G.; validation, Y.Z. and J.-a.L.; formal analysis, Z.H.; investigation, Z.H.; resources, Y.Z.; data curation, Y.Z.; writing—original draft preparation, Z.H.; writing—review and editing, J.-a.L. and Y.Z.; visualization, Y.Z.; supervision, L.W.; funding acquisition, Y.Z., K.Z. and J.-a.L. All authors have read and agreed to the published version of the manuscript.

Funding: This research was funded by National Nature Science Foundation of China (No. U2004164 and 51671175).

Institutional Review Board Statement: Not applicable.

Informed Consent Statement: Not applicable.

Data Availability Statement: The data presented in this study are available on request from the authors.

Conflicts of Interest: The authors declare no conflict of interest.

References

1. Roth, G.A.; Johnson, C.; Abajobir, A.; Abd-Allah, F.; Abera, S.F.; Abyu, G.; Ahmed, M.; Aksut, B.; Alam, T.; Alam, K.; et al. Regional, and National Burden of Cardiovascular Diseases for 10 Causes, 1990 to 2015. *J. Am. Coll. Cardiol.* **2017**, *70*, 1–25. [CrossRef] [PubMed]
2. Li, J. Advanced Biomaterials for Drug Delivery and Tissue Regeneration. *Curr. Drug Deliv.* **2021**, *18*, 834–835. [CrossRef] [PubMed]
3. Crimi, G.; Gritti, V.; Galiffa, V.A.; Scotti, V.; Leonardi, S.; Ferrario, M.; Ferlini, M.; De Ferrari, G.M.; Visconti, L.O.; Klersy, C. Drug eluting stents are superior to bare metal stents to reduce clinical outcome and stent-related complications in CKD patients, a systematic review, meta-analysis and network meta-analysis. *J. Interv. Cardiol.* **2018**, *31*, 319–329. [CrossRef] [PubMed]
4. Wei, S.; Xu, Y.; Wang, Z.; Li, M.; Sun, P.; Xie, B.; Xing, Y.; Bai, H.; Kan, Q.; Li, J.; et al. Hydrogel-coated needles prevent puncture site bleeding. *Acta Biomater.* **2021**, *128*, 305–313. [CrossRef] [PubMed]

5. Iakovou, I.; Schmidt, T.; Bonizzoni, E.; Ge, L.; Sangiorgi, G.M.; Stankovic, G.; Airoldi, F.; Chieffo, A.; Montorfano, M.; Carlino, M.; et al. Incidence, Predictors, and Outcome of Thrombosis After Successful Implantation of Drug-Eluting Stents. *JAMA* **2005**, *293*, 2126–2130. [CrossRef]
6. Yu, Y.; Zhu, S.; Hou, Y.; Li, J.; Guan, S. Sulfur contents in sulfonated hyaluronic acid direct the cardiovascular cells fate. *ACS Appl. Mater. Interfaces* **2020**, *12*, 46827–46836. [CrossRef]
7. Sketch, M.H.; Ball, M.; Rutherford, B.; Popma, J.J.; Russell, C.; Kereiakes, D.J.; on behalf of the Driver Investigators. Evaluation of the Medtronic (Driver) cobalt-chromium alloy coronary stent system. *Am. J. Cardiol.* **2005**, *95*, 8–12. [CrossRef]
8. Dong, H.; Li, D.; Mao, D.; Bai, N.; Chen, Y.; Li, Q. Enhanced performance of magnesium alloy for drug-eluting vascular scaffold application. *Appl. Surf. Sci.* **2018**, *435*, 320–328. [CrossRef]
9. Hou, Y.; Witte, F.; Li, J.; Guan, S. The increased ratio of Mg^{2+}/Ca^{2+} from degrading magnesium alloys directs macrophage fate for functionalized growth of endothelial cells. *Smart Mater. Med.* **2022**, *3*, 188–198. [CrossRef]
10. Witte, F.; Hort, N.; Vogt, C.; Cohen, S.; Kainer, K.U.; Willumeit, R.; Feyerabend, F. Degradable biomaterials based on magnesium corrosion. *Curr. Opin. Solid State Mater. Sci.* **2008**, *12*, 63–72. [CrossRef]
11. Liu, C.; Hou, Y.; Li, J. A mini review on biodegradable magnesium alloy vascular stent. *Adv. Mater. Lett.* **2020**, *11*, 20101563. [CrossRef]
12. Zhao, Z.; Moghadasian, M.H. Chemistry, natural sources, dietary intake and pharmacokinetic properties of ferulic acid: A review. *Food Chem.* **2008**, *109*, 691–702. [CrossRef] [PubMed]
13. Kumar, N.; Pruthi, V. Potential applications of ferulic acid from natural sources. *Biotechnol. Rep.* **2014**, *4*, 86–93. [CrossRef] [PubMed]
14. Sanduk, F.; Meng, Y.; Widera, D.; Kowalczyk, R.M.; Michael, N.; Kaur, A.; Yip, V.; Zulu, S.; Zavrou, I.; Hana, L.; et al. Enhanced anti-inflammatory potential of degradation resistant curcumin/ferulic acid eutectics embedded in triglyceride-based microemulsions. *J. Drug Deliv. Sci. Technol.* **2020**, *60*, 102067. [CrossRef]
15. Wang, B.H.; Ou-Yang, J.P. Pharmacological Actions of Sodium Ferulate in Cardiovascular System. *Cardiovasc. Drug Rev.* **2010**, *23*, 161–172. [CrossRef]
16. Lin, C.M.; Chiu, J.H.; Wu, I.H.; Wang, B.W.; Pan, C.M.; Chen, Y.H. Ferulic acid augments angiogenesis via VEGF, PDGF and HIF-1 alpha. *J. Nutr. Biochem.* **2010**, *21*, 627–633. [CrossRef]
17. Wang, J.; Yuan, Z.; Zhao, H.; Ju, D.; Chen, Y.; Chen, X.; Zhang, J. Ferulic acid promotes endothelial cells proliferation through up-regulating cyclin D1 and VEGF. *J. Ethnopharmacol.* **2011**, *137*, 992–997. [CrossRef]
18. Hou, Y.Z.; Yang, J.; Zhao, G.R.; Yuan, Y.J. Ferulic acid inhibits vascular smooth muscle cell proliferation induced by angiotensin II. *Eur. J. Pharmacol.* **2004**, *499*, 85–90. [CrossRef]
19. Zhang, E.; Shen, F. Blood compatibility of a ferulic acid (FA)-eluting PHBHHx system for biodegradable magnesium stent application. *Mater. Sci. Eng. C Mater. Biol. Appl.* **2015**, *52*, 37–45. [CrossRef]
20. Li, J.; Chen, L.; Zhang, X.; Guan, S. Enhancing biocompatibility and corrosion resistance of biodegradable Mg-Zn-Y-Nd alloy by preparing PDA/HA coating for potential application of cardiovascular biomaterials. *Mater. Sci. Eng. C* **2020**, *109*, 110607. [CrossRef]
21. Liang, J.; Cui, L.; Li, J.K.; Guan, S.; Zhang, K.; Li, J. Aloe vera: A medicinal plant used in skin wound healing. *Tissue Eng. Part B Rev.* **2021**, *27*, 455–474. [CrossRef] [PubMed]
22. Wei, S.; Li, J.; He, H.; Shu, C.; Dardik, A.; Bai, H. A three-layered hydrogel patch with hierarchy releasing of PLGA nanoparti-cle drugs decrease neointimal hyperplasia. *Smart Mater. Med.* **2022**, *3*, 139–147. [CrossRef]
23. Ho, C.C.; Ding, S.J. Structure, properties and applications of mussel-inspired polydopamine. *J. Biomed. Nanotechnol.* **2014**, *10*, 3063–3084. [CrossRef] [PubMed]
24. Chen, D.; Wang, R.; Huang, Z.; Wu, Y.; Zhang, Y.; Wu, G.; Li, D.; Guo, C.; Jiang, G.; Yu, S.; et al. Evolution processes of the corrosion behavior and structural characteristics of plasma electrolytic oxidation coatings on AZ31 magnesium alloy. *Appl. Surf. Sci.* **2018**, *434*, 326–335. [CrossRef]
25. Liu, X.; Zhen, Z.; Liu, J.; Xi, T.; Zheng, Y.; Guan, S.; Zheng, Y.; Cheng, Y. Multifunctional MgF_2/Polydopamine Coating on Mg Alloy for Vascular Stent Application. *J. Mater. Sci. Technol.* **2015**, *31*, 733–743. [CrossRef]
26. Tong, P.; Sheng, Y.; Hou, R.; Iqbal, M.; Chen, L.; Li, J. Recent progress on coatings of biomedical magnesium alloy. *Smart Mater. Med.* **2022**, *3*, 104–116. [CrossRef]
27. Yang, Y.X.; Fang, Z.; Liu, Y.H.; Hou, Y.C.; Wang, L.G.; Zhou, Y.F.; Zhu, S.J.; Zeng, R.C.; Zheng, Y.F.; Guan, S.K. Biodegradation, hemocompatibility and covalent bonding mechanism of electrografting polyethylacrylate coating on Mg alloy for cardiovascular stent. *J. Mater. Sci. Technol.* **2020**, *46*, 114–126. [CrossRef]
28. Huang, W.; Mei, D.; Liu, Y.; Wang, L.; Zhou, Y.; Zhu, S.; Guan, S. A simple approach for synthesizing polyglycolide coating on magnesium alloy. *Mater. Lett.* **2021**, *298*, 130046. [CrossRef]
29. Liu, J.; Zheng, B.; Wang, P.; Wang, X.; Zhang, B.; Shi, Q.; Xi, T.; Chen, M.; Guan, S. Enhanced in Vitro and in Vivo Performance of Mg-Zn-Y-Nd Alloy Achieved with APTES Pretreatment for Drug-Eluting Vascular Stent Application. *ACS Appl. Mater. Interfaces* **2016**, *8*, 17842–17858. [CrossRef]
30. Liu, J.; Xi, T. Enhanced Anti-corrosion Ability and Biocompatibility of PLGA Coatings on MgZnYNd Alloy by BTSE-APTES Pre-treatment for Cardiovascular Stent. *J. Mater. Sci. Technol.* **2016**, *32*, 845–857. [CrossRef]

31. Sheng, Y.; Yang, J.; Hou, R.; Chen, L.; Xu, J.; Liu, H.; Zhao, X.; Wang, X.; Zeng, R.; Li, W.; et al. Improved biocompatibility and degradation behavior of biodegradable Zn-1Mg by grafting zwitterionic phosphorylcholine chitosan (PCCs) coating on silane pre-modified surface. *Appl. Surf. Sci.* **2020**, *527*, 146914. [CrossRef]
32. Chen, Y.; Wan, G.; Wang, J.; Zhao, S.; Zhao, Y.; Huang, N. Covalent immobilization of phytic acid on Mg by alkaline pre-treatment: Corrosion and degradation behavior in phosphate buffered saline. *Corros. Sci.* **2013**, *75*, 280–286. [CrossRef]
33. Yang, Z.; Xiong, K.; Qi, P.; Yang, Y.; Tu, Q.; Wang, J.; Huang, N. Gallic acid tailoring surface functionalities of plasma-polymerized allylamine-coated 316L SS to selectively direct vascular endothelial and smooth muscle cell fate for enhanced endothelialization. *ACS Appl. Mater. Interfaces* **2014**, *6*, 2647–2656. [CrossRef]
34. Chen, L.; Li, J.; Chang, J.; Jin, S.; Wu, D.; Yan, H.; Wang, X.; Guan, S. Mg-Zn-Y-Nd coated with citric acid and dopamine by layer-by-layer self-assembly to improve surface biocompatibility. *Sci. China Technol. Sci.* **2018**, *61*, 1228–1237. [CrossRef]
35. Li, J.; Li, W.; Zou, D.; Kou, F.; Hou, Y.; Yasin, A.; Zhang, K. Comparison of conjugating chondroitin sulfate A and B on amine-rich surface: For deeper understanding on directing cardiovascular cells fate. *Compos. Part B Eng.* **2022**, *228*, 109430. [CrossRef]
36. Bai, H.; Sun, P.; Wu, H.; Wei, S.; Xie, B.; Wang, W.; Hou, Y.; Li, J.; Dardik, A.; Li, Z. The application of tissue-engineered fish swim bladder vascular graft. *Commun. Biol.* **2021**, *4*, 1153. [CrossRef] [PubMed]
37. Yu, Y.; Zhu, S.J.; Dong, H.T.; Zhang, X.Q.; Li, J.A.; Guan, S.K. A novel MgF_2/PDA/S-HA coating on the biodegradable ZE21B alloy for better multi-functions on cardiovascular application. *J. Magnes. Alloys* **2021**. [CrossRef]
38. Zou, D.; Li, J.; Kou, F.; Luo, X.; Yang, P. Reveal crucial subtype of natural chondroitin sulfate on the functionalized coatings for cardiovascular implants. *J. Mater. Sci. Technol.* **2021**, *91*, 67–77. [CrossRef]
39. Li, J.; Wang, S.; Sheng, Y.; Liu, C.; Xue, Z.; Tong, P.; Guan, S. Designing HA/PEI nanoparticle composite coating on biodegradable Mg-Zn-Y-Nd alloy to direct cardiovascular cells fate. *Smart Mater. Med.* **2021**, *2*, 124–136. [CrossRef]
40. Kou, F.; Liu, C.; Wang, L.; Yasin, A.; Li, J.; Guan, S. Fabrication of Citric Acid/RGD Multilayers on Mg-Zn-Y-Nd Alloy via Layer-by-Layer Self-Assembly for Promoting Surface Biocompatibility. *Adv. Mater. Interfaces* **2021**, *8*, 2002241. [CrossRef]
41. Wu, Y.; Chang, L.; Li, J.; Wang, L.; Guan, S. Conjugating heparin, REDV peptide and anti-CD34 to the silanic Mg-Zn-Y-Nd alloy for better endothelialization. *J. Biomater. Appl.* **2020**, *35*, 158–168. [CrossRef] [PubMed]
42. Xu, R.; Zhang, K.; Liang, J.; Gao, F.; Li, J.; Guan, F. Hyaluronic acid/polyethyleneimine nanoparticles loaded with copper ion and disulfiram for esophageal cancer. *Carbohydr. Polym.* **2021**, *261*, 117846. [CrossRef] [PubMed]
43. Wei, Z.; Tian, P.; Liu, X.; Zhou, B. In vitro degradation, hemolysis, and cytocompatibility of PEO/PLLA composite coating on biodegradable AZ31 alloy. *J. Biomed. Mater. Res. Part B* **2015**, *103B*, 342–354. [CrossRef] [PubMed]

metals

Article

Tailoring of Biodegradable Magnesium Alloy Surface with Schiff Base Coating via Electrostatic Spraying for Better Corrosion Resistance

Yulong Sheng [1], Ruiqing Hou [1], Changsheng Liu [1], Zhonghua Xue [1], Kun Zhang [2], Jingan Li [1,*] and Shaokang Guan [1,*]

1 School of Materials Science and Engineering, Zhengzhou University, 100 Science Road, Zhengzhou 450001, China; ylsheng@gs.zzu.edu.cn (Y.S.); rq_hou1009@hotmail.com (R.H.); lcszzu123@163.com (C.L.); 202012192013543@gs.zzu.edu.cn (Z.X.)
2 School of Life Science, Zhengzhou University, 100 Science Road, Zhengzhou 450001, China; zhangkun@zzu.edu.cn
* Correspondence: lijingan@zzu.edu.cn (J.L.); skguan@zzu.edu.cn (S.G.); Tel.: +86-185-3995-6211 (J.L.)

Abstract: In this study, three new Schiff bases were synthesized from paeonol and amino acids to prepare a compound Schiff base coating on the Mg-Zn-Y-Nd alloy (ZE21B alloy) surface by electrostatic spraying, and these three single Schiff base coatings were prepared on the ZE21B alloy as control. The results of SEM and XPS confirmed the successful preparation of the coating. Immersion tests and electrochemical tests showed that both the single coating and the compound coating significantly improved the corrosion resistance of ZE21B alloy, and the compound coating could play a synergistic corrosion inhibition effect, thus showing the best corrosion resistance.

Keywords: biodegradable magnesium alloy; synergistic corrosion inhibition effect; Schiff base coating; electrostatic spraying

Citation: Sheng, Y.; Hou, R.; Liu, C.; Xue, Z.; Zhang, K.; Li, J.; Guan, S. Tailoring of Biodegradable Magnesium Alloy Surface with Schiff Base Coating via Electrostatic Spraying for Better Corrosion Resistance. *Metals* **2022**, *12*, 471. https://doi.org/10.3390/met12030471

Academic Editor: Tullio Monetta

Received: 5 February 2022
Accepted: 7 March 2022
Published: 11 March 2022

1. Introduction

Magnesium (Mg) alloy has a broad development prospect because of its good electrical conductivity, thermal conductivity, low density, high specific strength, and suitable elastic modulus, and is known as the green engineering material in the new century [1–7]. In recent years, Mg alloy has attracted much attention in the field of biomedical materials due to its good biocompatibility and biodegradability, but its poor corrosion resistance is the main limiting factor for its clinical application. The biomedical Mg alloys have a strong tendency of local corrosion in the in vivo environment, which will seriously affect their mechanical properties and cannot give full play to their expected biological functions [8–14]. The application of a corrosion inhibitor is a common and efficient method to prevent and reduce the corrosion of metal materials. Therefore, tailoring of the Mg alloy surface with corrosion inhibitor may be an effective strategy to enhance the corrosion resistance [15–20].

Schiff base is a kind of organic corrosion inhibitor with the advantages of high efficiency, non-toxicity, easy preparation, easy storage, and low cost. Its molecular structure contains functional groups (C=N−) which can combine with metal ions such as Mg^{2+} and Zn^{2+} to form stable complexes, and it has great potential in the field of metal material corrosion protection [21–24]. Ma et al. synthesized three kinds of amino acid Schiff bases using paeonol and amino acids and the study showed that Schiff bases formed an intact and dense film layer on the surface of Mg alloys by complexation reaction, which significantly decreased the corrosion rate and improved the corrosion resistance of the alloys [25]. However, for small Mg alloy implant devices, the immersion method is not suitable for preparing Schiff base coating, because the conventional immersion time will damage the device structure due to excessive corrosion, and the shorter immersion time cannot promise a complete and dense coating formed on the surface. The electrostatic spraying technology provides inspiration for the Schiff base coating preparation benefit from its advantages of the shorter time and higher effectiveness [26–28].

In this study, three kinds of novel Schiff bases were synthesized and a compound coating composed of these three Schiff bases were prepared on the Mg-Zn-Y-Nd alloy (ZE21B alloy) surface by electrostatic spraying. This compound coating was anticipated to endow the ZE21B alloy stronger corrosion resistance.

2. Materials and Methods

2.1. Experimental Materials

The ZE21B alloy (Mg-2.0Zn-0.46Y-0.5Nd alloy) was independently developed in Henan Key Laboratory of Advanced Magnesium Alloy (Zhengzhou, China), and the cylindrical ZE21B alloy were cut into small discs (Φ 10 mm × 3 mm), then polished with silicon carbide sandpaper and diamond polishing agent, and then ultrasonic cleaned in acetone and anhydrous ethanol successively for 3 min, and dried.

2.2. Synthesis of Schiff Base

Paeonol (natural organic matter extracted from peony root bark) was reacted with three kinds of amino acids (lysine, glycine, and methionine) to synthesize Schiff base containing amino group (−RC=N−) as described in the previous study [25], and the actual yield of all three Schiff bases can be beyond 80.0% theoretical yield. The molecular structures of the three synthesized Schiff bases: lysine Schiff base (PCLys), glycine Schiff base (PCGly), and methionine Schiff base (PCMet) are shown in Figure 1.

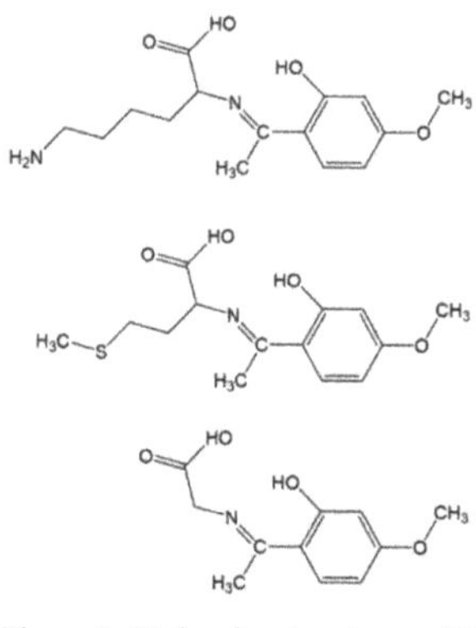

Figure 1. Molecular structures of Schiff base: PCLys, PCGly, and PCMet.

2.3. Preparation of Schiff Base Coatings on ZE21B Alloy

Schiff base coating was prepared on the surface of ZE21B alloy by electrospray technique. The synthesized Schiff base was dissolved in absolute ethanol and stirred vigorously at 1500 rpm for 36–48 h at a constant temperature of 30–40 °C using a magnetic heating stirrer (Hunan Saiwei Technology Co., Ltd, Changsha, China) to obtain a relatively homogeneous and stable Schiff base solution after being pumped into a dedicated syringe. Then, the Schiff base solution was dispersed into small droplets with very uniform fine microstructure using an applied high-voltage electrostatic field, and the small droplets were sprayed from the needle tip position and punched onto the ZE21B surface to form a uniform and dense layer of single Schiff base coating or compound Schiff base coating. In order to achieve a stable Taylor cone spray pattern for liquids sprayed from the needle tip position to form an ideal stable spray effect, the specific process parameters for electrospray were set as follows: spray distance of 35 mm, spray voltage of 7.0–9.0 kV, and spray flow rate of 10–20 µL/min (spray time 2 min). The concentrations of Schiff base solution used to prepare the single Schiff base coatings were designed as: PCLys = 0.05 M,

PCGly = 0.07 M, PCMet = 0.07 M, and the concentration ratio of the 3 Schiff bases was PCLys: PCGly: PCMet = 1:1:2.

2.4. Characterization of Schiff Base Coatings on ZE21B Alloy

The micro-morphologies of the PCLys-, PCGly-, PCMet-, compound Schiff base-coated ZE21B, and the bare ZE21B were analyzed by scanning electron microscopy (SEM, FEIQuanta200, Eindhoven, Holland) [29], and the element content and distribution on each sample were analyzed by energy dispersion spectrometer (EDS, FEIQuanta200, Eindhoven, Holland) [30]. The functional groups of PCLys, PCGly, PCMet, and compound coatings on ZE21B were detected by Fourier transform infrared spectroscopy (FTIR, Nicolet IS50, Thermo Fisher Scientific, Waltham, MA, USA) [31], and the element composition was examined by X-ray photoelectron spectroscopy (XPS, AXIS Supra, Kratos, Japan) [32]. The evaluation and validation of the data were carried out with the software CasaXPS (Version 2.3.24, Casa Software Ltd, Teignmouth, UK). Calibration of the spectra was done by adjusting the C1s signal to 284.5 eV.

2.5. Immersion Test

The non-working surface of the samples to be tested was sealed with silicone rubber before immersion experiments and weighed; The samples were transferred into a centrifuge tube and soaked with 25 mL RPMI 1640 cell culture medium (Art.No.:C3010-0500) and subsequently placed in a 37 °C constant temperature water bath for 1 day and 3 days. The pH value of each solution was measured every 24 h. At the 1st and 3rd day, the samples which retained the corrosion products were observed by SEM and EDS to investigate their pile up distribution and the deposition of calcium phosphorus salts, as well as analyze the degradation behavior of the coatings. The samples whose corrosion products were cleaned by ultrasonication in chromic acid solution (200 g/L CrO_3 + 10 g/L $AgNO_3$) for 2 min, sequentially cleaned with ultrapure water and 100% ethanol, and dried in oven at 37 °C. The weight loss was used to determine the average corrosion rate v and corrosion inhibition efficiency η (the calculation formula was shown as follow), and the corrosion of the sample surfaces was observed by SEM and EDS.

$$v = \frac{W_0 - W_1}{s \cdot t} \quad (1)$$

In the formula, W_0 and W_1 are the mass of the sample before and after soaking, s is the exposed area of the sample, and t is the soaking time. The formula for calculating the corrosion inhibition efficiency η is shown in Formula (2):

$$\eta = \frac{V_0 - V_1}{V_0} \times 100\% \quad (2)$$

In the formula, V_0 and V_1 are the average corrosion rates of bare ZE21B and Schiff base-coated ZE21B, respectively. The formula for calculating the degradation rate CR_w of the sample is shown in Formula (3):

$$CR_w = 3.65\frac{\Delta w}{D} \quad (3)$$

In the formula, CR_w is the average degradation rate, mm y^{-1}, Δw is the mass loss of the sample before and after immersion, and D is the material density.

2.6. Electrochemical Test

The electrochemical behavior of the sample in HBSS (Hank's balanced salt solution, pH 7.4) was investigated by using RST5200 electrochemical workstation. The three-electrode mode was adopted, platinum electrode was used as counter electrode, saturated calomel electrode was used as reference electrode, and the exposing surface of test sample is used as the working electrode. The formula of HBSS used is shown in Table 1.

Table 1. Formula of HBSS.

Component	Concentration (g/L)
NaCl	8.0
$MgSO_4 \cdot 7H_2O$	0.1
KCl	0.4
$MgCl_2 \cdot 6H_2O$	0.1
$CaCl_2$	0.14
$Na_2HPO_4 \cdot 12H_2O$	0.152
$KH_2PO_4 \cdot 3H_2O$	0.06
$C_6H_{12}O_6$	1.0
$NaHCO_3$	0.35

The potentiodynamic polarization measurements were conducted for the samples at a scanning rate of 1 mV/s from −1.7 to −1.0 V (vs.) open circuit potential (OCP) in HBSS after 1 h immersion. The obtained curves were fitted to obtain corrosion potential E_{corr} and corrosion current density I_{corr}, which were used to analyze the anti-corrosion properties of the coatings. Electrochemical impedance spectroscopy (EIS, Zhengzhou Shi Ruisi Instrument Technology Co., Ltd., Zhengzhou, China) was performed with the frequency range of 100,000–0.1 Hz at an open circuit potential after 30 min exposure in HBSS, and the alternating current voltage amplitude was 5 mV. The obtained EIS spectra were fitted by ZSimpWin software (Version 2.0, AMETEK Scientific Instruments, Minneapolis, MN, USA). The sum resistance (R_{sum}) of the samples for EIS-fitted results was evaluated to compare with the results obtained from polarization curves and immersion tests. The calculation formula of sum resistance R_{sum} is as follows:

$$R_{sum} = R_1 + R_2 + R_3 \tag{4}$$

R_1 represents the resistance of the Schiff base coating, R_2 represents the resistance of the corrosion product layer, and R_3 represents the charge transfer resistance.

3. Results and Discussion

Figure 2 displays the morphologies and element composition of PCLys-, PCGly-, PCMet-, and compound Schiff base-coated ZE21B samples, and the bare ZE21B. The ZE21B alloy showed a flat and uniform surface with many white secondary phase particles distributed on it, and EDS analysis results showed that the ZE21B contained a large amount of Mg (atomic percentage up to 95.77%) and a small amount of Zn, Y, and Nd on its surface. There were numerous long and short needle-like structures scattered disorderly on the surface of the four Schiff base coatings to form a layer of uniform and dense coating, and the discovery on the surface of N, which is a characteristic element in Schiff base, indicated that four coatings were successfully prepared on the ZE21B. In addition, four coatings also presented lower Mg ratios compared to the bare ZE21B surface.

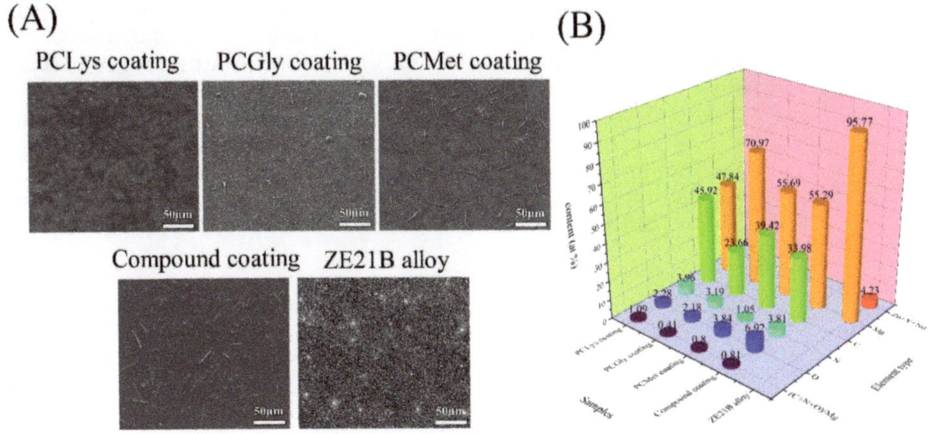

Figure 2. (**A**) SEM images of PCLys-, PCGly-, PCMet-, and compound Schiff base-coated ZE21B samples, and the bare ZE21B; (**B**) EDS element content histogram of each sample surface.

The immobilization of organic functional molecules will cause changes in the content of functional groups and elements on the ZE21B surface. In order to furtherly confirm the successful preparation of Schiff base coatings, each sample surface was tested and analyzed by FTIR (Figure 3A). It could be seen that the FTIR curve of ZE21B alloy was relatively flat, the intensity of characteristic absorption peaks was relatively weak; the reason may be that there is no special chemical molecular structure on the surface of the bare matrix, so rare absorption peaks could be detected. The curve of Schiff base coatings showed a new peak with relatively high intensity at 1605 cm^{-1} (PCLys and compound coatings), 1613 cm^{-1} (PCGly coating), 1597 cm^{-1} (PCMet coating), 1522 cm^{-1} (compound coating), and 1413 cm^{-1} (compound coating) corresponding to the stretching vibration absorption peak of the Cymene double bond, which was identified as the characteristic peak of the Schiff base molecule. Meanwhile, the Cymene double bond is also the key chemical group for the Schiff base molecule to show its corrosion inhibition function. In addition, an obvious absorption peak was observed near 3380 cm^{-1}, which was wide and blunt, and it belongs to the characteristic absorption peak of −OH.

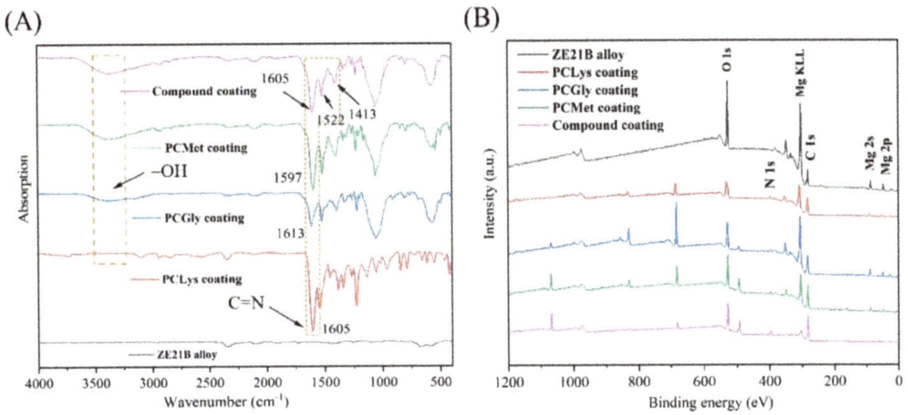

Figure 3. (**A**) FT-IR spectrum and (**B**) XPS full spectrum of PCLys-, PCGly-, PCMet-, and compound Schiff base-coated ZE21B samples, and the bare ZE21B.

The XPS full spectrum in Figure 3B presented the elements content on each surface: Compared with the four kinds of Schiff base coatings, the characteristic peaks of Mg KLL (eV), Mg 2s (eV), and Mg 2p (eV) in ZE21B alloy were more obvious, and the peak intensity was relatively higher. The Schiff base coatings presented obvious characteristic peaks of C 1s, N 1s, and O 1s with relatively high peak intensity due to their high content of C, N, and O elements.

Therefore, through the comparative analysis of the FTIR spectra and XPS full spectrum, it is sufficient to prove that the four kinds of Schiff base coatings were successfully prepared on the ZE21B alloy.

3.1. Immersion Test

The corrosion resistance properties of four kinds of Schiff base coatings on ZE21B alloy in RPMI 1640 cell culture medium were tested by static weight loss method. Tables 2 and 3 presented that all the Schiff base coatings can significantly reduce the corrosion rate of ZE21B alloy in cell culture medium, suggesting effective improvement of corrosion resistance, wherein compound coating had the higher corrosion resistance efficiency and lower corrosion rate compared to the other coatings. In addition, the pH value of the bare ZE21B alloy rapidly increased from the 24th h to the 72th h, indicating a typical alkaline environment, which is not conductive to cell growth, while the pH values of Schiff base coated ZE21B alloy kept in stable, suggesting a neutral or weakly alkaline environment, which is conductive to cell growth [33].

Table 2. Static weightlessness test for 24 h.

Samples	Corrosion Rate v ($mg\cdot cm^{-2}\cdot h^{-1}$)	Degradation Rate CR_w ($mm\cdot y^{-1}$)	Corrosion Resistance Efficiency η	pH
PCLys coating	0.0961	4.729	36.36%	7.09
PCGly coating	0.0794	3.908	47.40%	7.12
PCMet coating	0.0883	4.346	41.56%	7.03
Compound coating	0.0569	2.800	62.34%	7.48
ZE21B alloy	0.1511	7.436	-	7.88
Blank control		-		7.32

Table 3. Static weightlessness test for 72 h.

Samples	Corrosion Rate v ($mg\cdot cm^{-2}\cdot h^{-1}$)	Degradation Rate CR_w ($mm\cdot y^{-1}$)	Corrosion Resistance Efficiency η	pH
PCLys coating	0.0935	4.601	17.34%	7.42
PCGly coating	0.0736	3.622	34.97%	7.24
PCMet coating	0.0288	1.417	74.57%	7.69
Compound coating	0.0281	1.383	75.14%	7.81
ZE21B alloy	0.1131	5.566	-	8.21
Blank control		-		7.45

The SEM images in Figure 4A depict the morphology of each sample with corrosion products after immersed in the RPMI 1640 cell culture medium for 24 h and 72 h. At the 24th h, corrosion occurred in the local area of the bare ZE21B alloy surface, pitting holes with different sizes unevenly distributed on its surface, and many clearly visible cracks appeared, while the four kinds of Schiff base coatings exhibited smoother and more uniform surfaces after soaking for 24 h, and there were no obvious corrosion pits on any coated samples. After soaking for 72 h, the local corrosion on the bare ZE21B alloy surface was serious, the number of corrosion pits on the surface increased and was deepened, the cracks and corrosion area increased, but the surfaces of the four Schiff base coatings were still relatively flat, and slight pitting corrosion could be observed, and there was no trend of corrosion deepening. The EDS results showed that the corrosion products on all the surfaces were composed of P, Ca, O, C, N, and Mg elements (Figure 4B,C), which indicated there may be calcium phosphate, MgO, or $Mg(OH)_2$ accumulated on each surface.

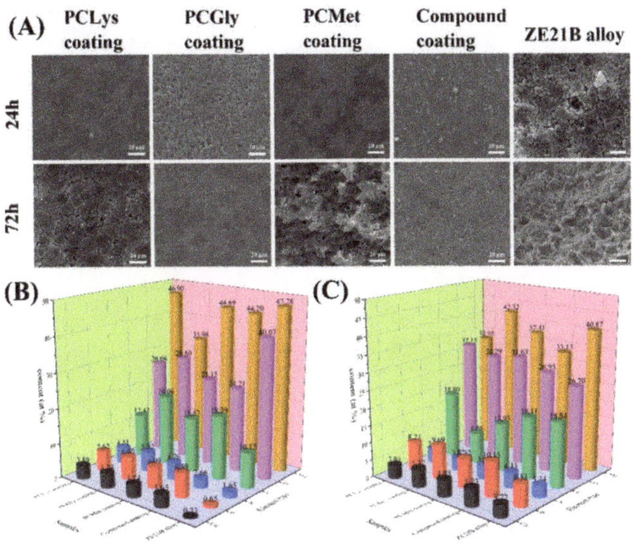

Figure 4. (**A**) SEM images of each sample with corrosion products after being immersed in the RPMI 1640 cell culture medium for 24 h and 72 h; EDS results of each sample with corrosion products after immersed in the RPMI 1640 cell culture medium for (**B**) 24 h and (**C**) 72 h.

Figure 5A depicts the morphology of each sample without corrosion products after being immersed in the RPMI 1640 cell culture medium for 24 h and 72 h. After immersing for 24 h, it could be seen that there were a large number of pitting holes on the ZE21B alloy surface and the corrosion cracks were clearly visible. Up to 72 h, the ZE21B alloy surface seriously cracked and became uneven, suggesting intensified local corrosion, and the cracks even continued to extend to the entire surface, seriously damaging the Mg alloy. By comparison, the Schiff base-coated samples kept their surfaces smooth and uniform during the whole period, and there were only few and very small corrosion pits and cracks sparsely distributed on the surface. The EDS results (Figure 5B,C) showed that although all the samples presented higher Mg element and lower C, O, and N elements, the compound Schiff base coating still had higher N element compared to the other surfaces, suggesting more Schiff base molecules, which may be the reason of its stronger corrosion resistance.

3.2. *The Inhibition Mechanism of Schiff-Base Coating*

The magnesium alloy reacts in the corrosion solution and Mg^{2+} and H_2 are produced. When the Schiff base coating is destroyed by the corrosive medium, the released Schiff base molecule can be reacted with the Mg^{2+} released from magnesium alloy, and the reaction process includes the deprotonization reaction and complexation reaction of the Schiff base molecule, such as in Figure 6. The generated Schiff base complex is difficult to dissolve in water, which can stably adsorb the surface of the magnesium alloy, and form a dense uniform protective film layer together with the corrosion product such as MgO, $Mg(OH)_2$ to prevent corrosion. The medium further contacts the substrate to contribute to the substrate, thereby generating the effect of alleviating corrosion, reducing the corrosion rate of the alloy, and greatly increasing its corrosion resistance [25].

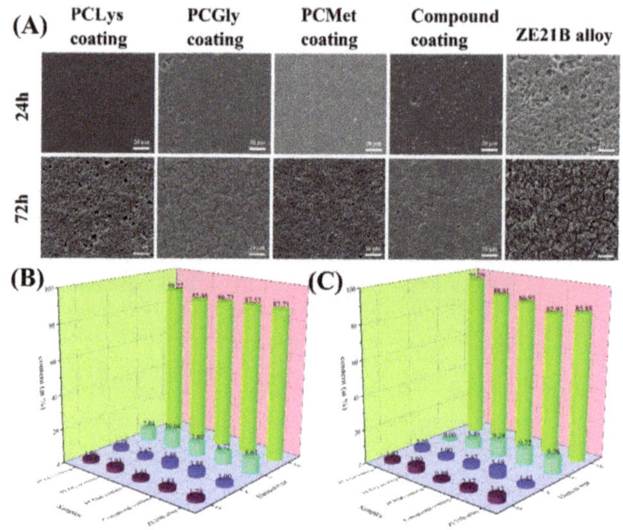

Figure 5. (**A**) SEM images of each sample without corrosion products after being immersed in the RPMI 1640 cell culture medium for 24 h and 72 h; EDS results of each sample with corrosion products after immersed in the RPMI 1640 cell culture medium for (**B**) 24 h and (**C**) 72 h.

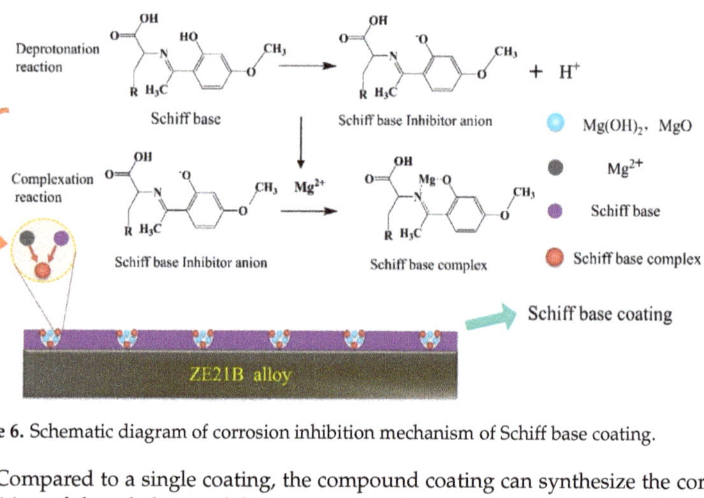

Figure 6. Schematic diagram of corrosion inhibition mechanism of Schiff base coating.

Compared to a single coating, the compound coating can synthesize the corrosion inhibition of three bakers, exhibiting a synergistic corrosion resistance, which exhibits stronger corrosion resistance. The mechanism is as follows: when single Schiff base coating is corroded, the formed protective film is not very dense, the erosion effect of the barrier corrosion medium is limited, and the synergistic effect of the three Schiff base molecules has strengthened the adsorption effect of Schiff base in magnesium alloy surface. The formed protective film layer is relatively denser, and can better block the corrosion medium erosion matrix and better protect the magnesium alloy, exhibiting higher corrosion inhibition efficiency.

3.3. Electrochemical Test

The self-corrosion potential (E_{corr}) and corrosion current density (I_{corr}) of the samples can be calculated from the polarization curves, as shown in Table 4. As can be seen from Figure 7A and Table 4, the E_{corr} value of ZE21B alloy is the lowest, −1.501 V, and the self-corrosion potential of each coating sample is significantly positive, which indicates that the corrosion tendency of each coating sample is higher than that of the single coating. The smaller the matrix, the more difficult it is to corrode. The I_{corr} value of ZE21B alloy is the highest, reaching 5.56 μA/cm^2. The current density of each coating sample has been significantly reduced, the corrosion rate has been greatly slowed down, and the corrosion resistance has been significantly improved. At the same time, it can be seen from Figure 7A that the shape of the polarization curves of different types of Schiff base coating samples did not change significantly, indicating that the corrosion inhibition mechanism of the Schiff base coating is not affected by the type of coating. Therefore, from the analysis results of the polarization curve, it can be seen that the Schiff base coating can significantly inhibit the corrosion of HBSS corrosion medium to magnesium alloys, and can play a better protective effect on ZE21B alloy, greatly improving its corrosion resistance and corrosion inhibition effect. The order from strong to weak is compound coating > PCMet coating > PCLys coating > PCGly coating, which is not much different from the results of the static weightless test.

Table 4. Electrochemical parameters.

Samples	*Ecorr* (V)	*Icorr* (μA/cm^2)
PCLys coating	−1.468	0.507
PCGly coating	−1.448	0.683
PCMet coating	−1.440	0.466
Compound coating	−1.427	0.440
ZE21B alloy	−1.501	5.560

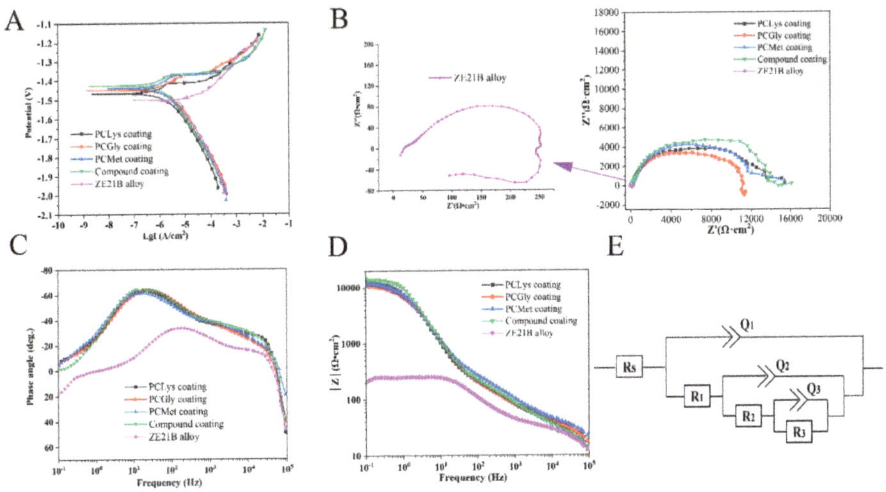

Figure 7. (**A**) Polarization curve diagram, (**B**) EIS spectrum diagram, (**C**) phase angle—bode diagram, (**D**) impedance modulus—bode diagram, (**E**) equivalent circuit diagram.

In order to study the corrosion resistance mechanism of different coatings in HBSS in detail, electrochemical impedance spectroscopy (EIS) was used to track the performance of the surface films of different samples. The EIS spectrum data can be analyzed with the equivalent circuit in Figure 7E, and the corresponding EIS spectrum can be fitted with ZSimpWin software. The relevant parameters are shown in Table 5. The equivalent circuit model R(Q(R(Q(R(QR))))) is more suitable for simulating organic coatings, where R_s represents the resistance of the HBSS between the platinum electrode and the reference electrode, R_1 and Q_1 represent the resistance and capacitance of the Schiff base coating, respectively, R_2 and Q_2 represent the resistance and capacitance of the corrosion product layer, respectively, R_3 and Q_3 represent the charge transfer resistance and capacitance, respectively, and n_1, n_2, and n_3 are the capacitance coefficients associated with Q_1, Q_2, and Q_3, respectively. Figure 7B is the Nyquist diagram of different samples in HBSS. It can be seen from the figure that the Nyquist diagram is composed of three semicircular arcs in the high, medium, and low frequency region. The incomplete impedance arc is due to the corrosion product film formed on the surface of the magnesium alloy. It is caused by various reasons such as the electrochemical reaction of the layer and the surface of the electrode. There are three capacitive reactance arcs in the EIS spectrum of the bare substrate and the coated sample, of which the high frequency is the capacitive reactance arc related to the Schiff base film on the surface of the sample, and the intermediate frequency is the capacitive reactance related to the corrosion product film. Anti-arcing, at low frequencies is the capacitive anti-arcing associated with the Faraday charge transfer resistance. Compared with the bare substrate, the impedance arc shape of the coated sample did not change significantly, but the impedance arc radius increased significantly, indicating that the corrosion product film on the surface of the sample was more complete and dense, which increased the resistance to charge transfer on the surface of the magnesium alloy, the stability of the corrosion product film layer improved, the corrosion resistance greatly improved, the electrochemical reaction rate greatly reduced, the corrosion of the magnesium alloy was significantly inhibited, and the corrosion resistance of the alloy significantly improved. The composite coating sample had the best corrosion resistance. The composite coating can exert the effect of synergistic corrosion inhibition and greatly improve the corrosion resistance of the coating, which is basically consistent with the results of the polarization curve.

Table 5. Equivalent circuit fitting parameters.

Samples	PCLys Coating	PCGly Coating	PCMet Coating	Compound Coating	ZE21B Alloy
R_s ($\Omega\cdot cm^2$)	26.1	30.7	31.0	28.4	23.3
R_1 ($\Omega\cdot cm^2$)	84.4	64.7	233	16,580	24.3
Q-Y_1 ($cm^2\cdot s^n\cdot\Omega$)	0	0	0	0	0
n_1	0.95	1	0.77	0.73	1
R_2 ($\Omega\cdot cm^2$)	12,660	11,000	12,430	1.3	77.3
Q-Y_2 ($cm^2\cdot s^n\cdot\Omega$)	0	0	0	0	0
n_2	0.80	0.78	0.83	0.78	1
R_3 ($\Omega\cdot cm^2$)	3037	0.7	3500	53.5	115.8
Q-Y_3 ($cm^2\cdot s^n\cdot\Omega$)	0	0.71	0	0	0
n_3	1	0.8	1	0.67	1
R_{sum} ($\Omega\cdot cm^2$)	15,781.4	11,065.4	16,163	16,634.8	217.4

Figure 7C,D are the Bode-phase angle diagrams and Bode-impedance modulus diagrams of different samples in HBSS. From Figure 7C, it can be found that the phase angle curves of each coating sample appear. There are three peaks, which correspond to three time constants, one appears in the high frequency range, which can be attributed to the physical barrier properties of the Schiff base coating, and the other appears in the intermediate frequency region, which is caused by the corrosion product film layer, while the low frequency region is generated by the electric double layer capacitance and charge transfer resistance at the interface between the magnesium alloy substrate and the coating. It can be seen from Table 5 that the R_{sum} of the composite coating sample is the largest, which further shows that the corrosion resistance of the composite coating sample is the best, which is basically consistent with the results of the EIS spectrum and polarization curve.

4. Conclusions

In this study, the PCLys coating, PCGly coating, PCMet coating, and compound coating composed of these three Schiff base molecules were successfully prepared and applied onto the ZE21B alloy surface by electrostatic spraying. All the Schiff base coatings have a certain degree of corrosion resistance and can show different degrees of corrosion inhibition efficiency, and the corrosion inhibition efficiency of the composite coating is significantly better than that of the single coating, showing relatively good corrosion resistance performance. Compared to the single Schiff base coating, the compound coating can play a synergistic corrosion inhibition effect and better protect the magnesium alloy, so it is possible to show better corrosion resistance.

Author Contributions: Conceptualization, Y.S. and J.L.; methodology, R.H.; software, Y.S.; validation, Y.S., C.L. and Z.X.; formal analysis, Y.S. and K.Z.; investigation, Y.S.; resources, S.G.; data curation, Y.S.; writing—original draft preparation, Y.S.; writing—review and editing, J.L. and K.Z.; visualization, J.L.; supervision, S.G.; project administration, J.L. and K.Z.; funding acquisition, S.G. All authors have read and agreed to the published version of the manuscript.

Funding: This research was funded by the Key Projects of the National Key Research and Development Program of China, grant number 2018YFC1106703.

Institutional Review Board Statement: Not applicable.

Informed Consent Statement: Not applicable.

Data Availability Statement: The data presented in this study are available on request from the corresponding author.

Conflicts of Interest: The authors declare no conflict of interest.

References

1. Maqbool, A.; Khan, N.Z.; Siddiquee, A.N. Towards Mg Based Light Materials of Future: Properties, Applications, Problems, and Their Mitigation. *J. Manuf. Sci. Eng.* **2022**, *144*, 030801. [CrossRef]
2. Tan, J.; Ramakrishna, S. Applications of Magnesium and Its Alloys: A Review. *Appl. Sci.* **2021**, *11*, 6861. [CrossRef]
3. Song, J.F.; She, J.; Chen, D.L.; Pan, F.S. Latest research advances on magnesium and magnesium alloys worldwide. *J. Magnes. Alloy.* **2020**, *8*, 1–41. [CrossRef]
4. Li, S.B.; Yang, X.Y.; Hou, J.T.; Du, W.B. A review on thermal conductivity of magnesium and its alloys. *J. Magnes. Alloy.* **2020**, *8*, 78–90. [CrossRef]
5. Zeng, Z.R.; Stanford, N.; Davies, C.H.J.; Nie, J.F.; Birbilis, N. Magnesium extrusion alloys: A review of developments and prospects. *Int. Mater. Rev.* **2019**, *64*, 27–62. [CrossRef]
6. Chen, J.X.; Tan, L.L.; Yu, X.M.; Etim, I.P.; Ibrahim, M.; Yang, K. Mechanical properties of magnesium alloys for medical application: A review. *J. Mech. Behav. Biomed.* **2018**, *87*, 68–79. [CrossRef]
7. Jayasathyakawin, S.; Ravichandran, M.; Baskar, N.; Chairman, C.A.; Balasundaram, R. Mechanical properties and applications of Magnesium alloy—Review. *Mater. Today Proc.* **2020**, *27*, 909–913. [CrossRef]
8. Xi, T.F.; Wei, L.N.; Liu, J.; Liu, X.L.; Zhen, Z.; Zheng, Y.F. Research Progress in Bioresorbable Magnesium Scaffolds. *Acta Metall. Sin.* **2017**, *53*, 1153–1167. [CrossRef]
9. Zheng, Y.F.; Yang, H.T. Research Progress in Biodegradable Metals for Stent Application. *Acta Metall. Sin.* **2017**, *53*, 1227–1237. [CrossRef]
10. Liu, Y.X.; Zhu, S.; Han, B.Y. Research progress in anodic hydrogen evolution of magnesium electrochemistry corrosion. *Cailiao Gongcheng* **2020**, *48*, 17–27. [CrossRef]
11. Zhang, Z.Q.; Yang, Y.X.; Li, J.A.; Zeng, R.C.; Guan, S.K. Advances in coatings on magnesium alloys for cardiovascular stents—A review. *Bioact. Mater.* **2021**, *6*, 4729–4757. [CrossRef]
12. Liu, Y.; Lu, B.H.; Cai, Z.X. Recent Progress on Mg- and Zn-Based Alloys for Biodegradable Vascular Stent Applications. *J. Nanomater.* **2019**, *2019*, 1310792. [CrossRef]

13. Esmaily, M.; Svensson, J.E.; Fajardo, S.; Birbilis, N.; Frankel, G.S.; Virtanen, S.; Arrabal, R.; Thomas, S.; Johansson, L.G. Fundamentals and advances in magnesium alloy corrosion. *Prog. Mater. Sci.* **2017**, *89*, 92–193. [CrossRef]
14. Mei, D.; Lamaka, S.V.; Lu, X.P.; Zheludkevich, M.L. Selecting medium for corrosion testing of bioabsorbable magnesium and other metals—A critical review. *Corros. Sci.* **2020**, *171*, 108722. [CrossRef]
15. Hu, T.; Ouyang, Y.J.; Xie, Z.H.; Wu, L. One-pot scalable in situ growth of highly corrosion-resistant MgAl-LDH/MBT composite coating on magnesium alloy under mild conditions. *J. Mater. Sci. Technol.* **2021**, *92*, 225–235. [CrossRef]
16. Kharitonov, D.S.; Zimowska, M.; Ryl, J.; Zielinski, A.; Osipenko, M.A.; Adamiec, J.; Wrzesinska, A.; Claesson, P.M.; Kurilo, I.I. Aqueous molybdate provides effective corrosion inhibition of WE43 magnesium alloy in sodium chloride solutions. *Corros. Sci.* **2021**, *190*, 109664. [CrossRef]
17. Liu, S.Q.; Li, Z.X.; Yu, Q.L.; Qi, Y.M.; Peng, Z.J.; Liang, J. Dual self-healing composite coating on magnesium alloys for corrosion protection. *Chem. Eng. J.* **2021**, *424*, 130551. [CrossRef]
18. Song, Y.W.; Liu, D.; Tang, W.N.; Dong, K.H.; Shan, D.Y.; Han, E.H. Comparison of the corrosion behavior of AM60 Mg alloy with and without self-healing coating in atmospheric environment. *J. Magnes. Alloy.* **2021**, *9*, 1220–1232. [CrossRef]
19. Lu, X.P.; Li, Y.; Ju, P.F.; Chen, Y.; Yang, J.S.; Qian, K.; Zhang, T.; Wang, F.H. Unveiling the inhibition mechanism of an effective inhibitor for AZ91 Mg alloy. *Corros. Sci.* **2019**, *148*, 264–271. [CrossRef]
20. Shin, Y.; Cho, K. Corrosion Behavior and Inhibition Studies of AZ31B Magnesium Alloy With and Without Cl- in the Alkaline Electrolytes in Addition with Various Inhibitor Additives. *Corros. Sci. Technol.* **2019**, *18*, 243–252. [CrossRef]
21. Saha, S.K.; Dutta, A.; Ghosh, P.; Sukul, D.; Banerjee, P. Novel Schiff-base molecules as efficient corrosion inhibitors for mild steel surface in 1 M HCl medium: Experimental and theoretical approach. *Phys. Chem. Chem. Phys.* **2016**, *18*, 17898–17911. [CrossRef]
22. Seifzadeh, D.; Bezaatpour, A.; Shamkhali, A.N.; Basharnavaz, H. Experimental and Theoretical Studies to Examine the Inhibition Effect of a Schiff Base against Magnesium Corrosion. *Trans. Indian Inst. Met.* **2016**, *69*, 1545–1555. [CrossRef]
23. Guo, Y.H.; Yang, S.; Feng, W.; Li, Y.; Cheng, Y. Eelectrochemical Study of Inhibition Effect of a Schiff Base towards Magnesium Alloy Corrosion. *Int. J. Electrochem. Sci.* **2016**, *11*, 6043–6051. [CrossRef]
24. Li, W.; Su, Y.; Ma, L.; Zhu, S.; Zheng, Y.; Guan, S. Sol-gel coating loaded with inhibitor on ZE21B Mg alloy for improving corrosion resistance and endothelialization aiming at potential cardiovascular application. *Colloids Surf. B Biointerfaces* **2021**, *207*, 111993. [CrossRef]
25. Ma, L.; Li, W.; Zhu, S.; Wang, L.; Guan, S. Corrosion inhibition of Schiff bases for Mg-Zn-Y-Nd alloy in normal saline: Experimental and theoretical investigations. *Corros. Sci.* **2021**, *184*, 109268. [CrossRef]
26. Müller, V.; Balvay, S.; Gaillard, C.; Tadier, S.; Gremillard, L.; Djurado, E. One-step fabrication of single-phase hydroxyapatite coatings on Ti-alloy implants by electrostatic spray deposition: From microstructural investigation to in vitro studies. *Surf. Coat. Technol.* **2021**, *427*, 127805. [CrossRef]
27. Chang, L.; Li, X.; Tang, X.; Zhang, H.; He, D.; Wang, Y.; Zhao, J.; Li, J.; Wang, J.; Zhu, S.; et al. Micro-patterned hydroxyapatite/silk fibroin coatings on Mg-Zn-Y-Nd-Zr alloys for better corrosion resistance and cell behavior guidance. *Front. Mater. Sci.* **2020**, *14*, 413–425. [CrossRef]
28. Jia, Z.Y.; Ma, C.Y.; Zhang, H.B. PLGA Coatings and PLGA Drug-Loading Coatings for Cardiac Stent Samples: Degradation Characteristics and Blood Compatibility. *Coatings* **2021**, *11*, 1427. [CrossRef]
29. Li, J.; Li, W.; Zou, D.; Kou, F.; Hou, Y.C.; Yasin, A.; Zhang, K. Comparison of conjugating chondroitin sulfate A and B on amine-rich surface: For deeper understanding on directing cardiovascular cells fate. *Compos. Part B Eng.* **2022**, *228*, 109430. [CrossRef]
30. Li, J.A.; Chen, L.; Zhang, X.Q.; Guan, S.K. Enhancing biocompatibility and corrosion resistance of biodegradable Mg-Zn-Y-Nd alloy by preparing PDA/HA coating for potential application of cardiovascular biomaterials. *Mater. Sci. Eng. C* **2020**, *109*, 110607. [CrossRef]
31. Yu, Y.; Zhu, S.; Hou, Y.; Li, J.; Guan, S. Sulfur Contents in Sulfonated Hyaluronic Acid Direct the Cardiovascular Cells Fate. *ACS Appl. Mater. Interfaces* **2020**, *12*, 46827–46836. [CrossRef] [PubMed]
32. Zou, D.; Li, J.A.; Kou, F.; Luo, X.; Yang, P. Reveal crucial subtype of natural chondroitin sulfate on the functionalized coatings for cardiovascular implants. *J. Mater. Sci. Technol.* **2021**, *91*, 67–77. [CrossRef]
33. Yu, Y.; Zhu, S.J.; Dong, H.T.; Zhang, X.Q.; Li, J.A.; Guan, S.K. A novel MgF2/PDA/S-HA coating on the bio-degradable ZE21B alloy for better multi-functions on cardiovascular application. *J. Magnes. Alloy.* **2021**, in press. [CrossRef]

metals

Article

Influence of the Manufacturing Process on the Corrosion and Mechanical Behavior of Esophageal Stents

Dino Alferi [1], Jaroslav Fojt [2,*], Eva Kristianova [2], Derek W. Edwards [3,4] and Hans-Ulrich Laasch [3,4]

1 Faculty of Medicine in Hradec Kralove, Charles University, 500 03 Hradec Kralove, Czech Republic; alferid@lfhk.cuni.cz
2 Department of Metals and Corrosion Engineering, University of Chemistry and Technology Prague, 166 28 Prague, Czech Republic; eva.kristianova@vscht.cz
3 Minnova Medical Foundation CIC, Wilmslow SK9 1HY, UK; dwe@minnova.uk (D.W.E.); hul@minnova.uk (H.-U.L.)
4 Department of Radiology, The Christie NHS Foundation Trust, Manchester M20 4BX, UK
* Correspondence: fojtj@vscht.cz

Abstract: Esophageal nitinol stents are an established method for treating swallowing difficulties caused by obstructing cancer. This raesearch investigates the influence of different qualities of raw metal alloys in combination with production technology on corrosion resistance in standardized simulated gastric fluid (SGF). Four different international stent manufacturers produced samples of their standard stents from nitinol sourced from three different alloy manufacturers. The stents were subjected to a 6-week immersion in SGF. During the immersion, the surface was studied at specified intervals using microscopy. The surface of the samples was also studied by X-ray Photoelectron Spectroscopy and after immersion the released ions were analyzed. Results demonstrated that both raw material and certain steps in the manufacturing process negatively affect corrosion resistance. Analysis of the SGF showed that the amount of nickel released is proportional to the degree of corrosion attack. Finally, current accepted standard test methods are inadequate for assessing susceptibility to corrosion by gastric acid and should take the low pH of the implanted environment into account. Conversely, certain measures in the manufacturing process are able to reduce the impact of the base material on corrosion susceptibility.

Keywords: biomaterials; shape memory alloys; titanium alloys; corrosion; characterization

Citation: Alferi, D.; Fojt, J.; Kristianova, E.; Edwards, D.W.; Laasch, H.-U. Influence of the Manufacturing Process on the Corrosion and Mechanical Behavior of Esophageal Stents. *Metals* **2023**, *13*, 1542. https://doi.org/10.3390/met13091542

Received: 19 July 2023
Revised: 25 August 2023
Accepted: 29 August 2023
Published: 1 September 2023

1. Introduction

The first clinical use of esophageal stents was reported in the 1960s in the treatment of malignant dysphagia [1]. Today, many different stent designs with different parameters are available and their mechanical properties including radial and expansion force, axial rigidity, and conformability vary greatly, depending on the stent skeleton design, as well as the material and method of covering [2,3]. Curiously, despite self-expanding metal stents (SEMS) having been in clinical practice for decades, the understanding of the clinical need and the ideal test methods for stents are poorly understood and still undefined [4].

At present, SEMS for the gastro-intestinal tract are almost universally made of "nitinol", a nickel-titanium super-alloy [5]. Stents are very often covered with polymers to prevent cancer or reactive granulation tissue from growing into the stent [6]. If the nitinol stent is placed in the esophagus for a long time, the implant may fail due to the corrosive environment of the human body in combination with material fatigue [7]. This applies particularly to stents placed partially or wholly within the stomach, where they are exposed to the low pH of hydrochloric acid contained in gastric secretions. In the case of esophageal carcinoma, 2/3 of tumors occur at the gastro-esophageal junction, requiring the lower end of the stent to be placed in the stomach. Nitinol is an equimolar alloy of titanium and nickel. It exhibits shape memory and superelasticity, as well as acceptable biocompatibility, which are the main reasons this material is used in the production of stents. Mechanical properties and corrosion resistance depend on the phase composition. The phase composition of input material shall be as homogenous as possible. The presence of other phases and impurities may adversely affect the above properties [8–11]. The corrosion resistance of nitinol is based on the formation of a uniform thin layer of TiO_2. This layer is a uniform barrier layer for corrosion prevention [7,12]. In order for the stent to acquire shape-memory with regards to its 3D-configuration, it must be heat-treated, usually with temperatures around 500 °C. Heat treatment, however, affects the thickness of the TiO_2 layer. The thickness of the oxide layer increases with the heat treatment temperature [13], as does the fatigue resistance [14]. On the other hand, if the TiO_2 layer is too thick, it may have an increased number of defects. Such a defective layer loses its corrosion-protective effects [15,16]. The thickness of the oxide layer can be changed by subsequent surface treatment, e.g., chemical polishing, electropolishing, and plasma electrolytic oxidation (PEO) [17]. Both methods reduce the oxide layer formed by the heat treatment [18].

The application of the surface treatment also improves the fatigue life, as it reduces the amount of surface defects [14]. The mechanical stress to which the stent material is subjected has a significant effect on the corrosion resistance. Mechanical stress causes the oxide layer to crack, exposing the unoxidized metal to the corrosive environment and accelerating the corrosion process [19]. The above-mentioned processes can lead to breakage of the nitinol wires, which can lead to stent failure. And thus, to the deterioration of the patient's condition. Several cases have been reported in clinical practice where partial or complete fracture of esophageal nitinol stents have occurred [20–23].

Another risk associated with corrosion of nitinol stents is exposure of the surrounding tissues to higher doses of nickel, which is released into the body during the corrosion process. A poor surface treatment causes a local increase in nickel concentration [24], which, in vascular animal experiments, can cause inflammation and stenosis of stented arteries [25].

In the production of nitinol stents, the quality of the raw input material on corrosion resistance plays a key role. Although suppliers are bound by strict standards, there is still space for minor variations, in particular the amount of inclusions, the state of the surface (oxide layer thickness), and the final processing of the wire (e.g., etching). At present, it is possible to produce stents from wires with different chemical compositions with different history of heat treatment and surface quality. Different surface qualities show different corrosion resistance [26].

The aim of this study is to determine the effect of different quality of input materials in combination with different production processes on the corrosion resistance of esophageal nitinol stents in a simulated gastric environment.

2. Materials and Methods

Experiments were conducted on nitinol esophageal stents made from 4 different manufacturers (A to D) by standard manufacturing process of each manufacturer. Each manufacturer was asked to supply stent samples of approximately 25 mm × 100 mm from batches of 0.25 mm nitinol wire acquired from three different wire suppliers (1 to 3). Because testing focused on the corrosion resistance of the stent skeleton, the samples were not provided with a polymer or other covering that can change the corrosion resistance.

The surface of the samples was studied by stereomicroscope Olympus DF Plapo 1 X (Olympus, Tokyo, Japan) and by scanning electron microscope (SEM) Tescan Vega3 LMU (Tescan, Brno, Czech Republic) before, during, and after the immersion in simulated gastric fluid (SGF). The composition of the raw wires was verified by Energy Dispersive Spectrometer (EDS) OXFORD Instruments INCA 350 (Abingdon, UK).

The chemical composition of the surface layer before and after immersion in SGF was analyzed by XPS. An ESCAProbeP (Omicron Nanotechnology Ltd., London, UK) was used for analysis. An aluminium anode with an excitation energy of 1486.7 eV was used as the

X-ray source. CasaXPS software 2.3.15 (Casa Software Ltd., Devon, IK)was used to evaluate the spectra.

The calculation of the oxide layer thickness was based on the free flight path of the electron between two inelastic collisions [27,28]. These calculated thicknesses are only approximate; only pure oxides were considered in the calculations, which, however, do not occur on the examined samples; here, they are always mixtures.

The concentration of the released ions within the SGF after 6-week immersion was measured by Atomic Absorption Spectroscopy (GBC 932Plus, Dandenong, Australia).

Most stent manufacturers test corrosion susceptibility according to the generic ASTM standard F-2129, which was originally aimed at biliary stents. However, the high pH of this does not reflect the hostile environment of esophago-gastric stents. Therefore, simulated gastric fluid prepared according to the ASTM 2528-06 standard [29] for gastrostomy tubes was chosen for testing, to reproduce the low pH and components of gastric juice encountered in the lower esophagus due to stent-induced reflux [30,31], or because the stent may need to be placed with its lower end within the stomach [32].

Table 1 shows the composition of SGF. All used substances were in p.a. quality. The value of pH was 1.2. The pH of the solution was also measured after immersion. Immersion was conducted at 37 °C.

Table 1. Composition of SGF.

Component	Concentration
NaCl	2 g/L
Pepsin	3.2 g/L
HCl	7 mL/L

During immersion, the stents were withdrawn from the SGF at the regular intervals (after 3 days of immersion and then weekly up to 6 weeks) to conduct analysis (optical microscopy, SEM, radial forces measurement), rinsed with ethanol and distilled water, and returned to SGF. This procedure was previously tested and compared with non-interrupted immersion. There were no differences in the results in both procedures.

The electrochemical measurement was realized in the standard three electrode set up with silver-silver chloride reference electrode (SSCE) and glassy carbon counter electrode. The measurement consists of the 1 h stabilization of the open circuit potential (OCP) followed by 1 mV/s polarization in the range ±20 mV/OCP. The polarization resistance, which is reciprocal to the corrosion rate, was then evaluated. The Reference600 (Gamry, Warminster, PA, USA) potentiostat was used. The measurement was realized on the knitted out wires from individual stents. The exposed area was defined by galvanic tape 3M 470 (3M, St. Paul, Minnesota, USA), which has sufficient adhesion to prevent crevice corrosion and to eliminate the influence of the free end of the wire on the results (the surface state at the shear point is different from the rest of the surface).

Radial resistive force (RRF) is the force exerted by the pressure-resistant stent against compression. Chronic outward force (COF) is the force the stent exerts on the surrounding environment during expansion. Forces were measured with a dedicated TTR2 radial force measuring instrument (Blockwise, Tempe, AZ, USA). Measurement was conducted at a temperature of 37 °C. Parameters of measurement are listed in Table 2.

Two manufacturers supplied straight stents; two supplied stents with larger ends (flared and dog-bone shaped)

The initial diameter in the case of measurement of samples from manufacturer D was 28 mm. This is due to the fact that the stent ends were of a larger diameter than the trunk. To accommodate this, an initial diameter of the compression tool was chosen so that there was no deformation of the stent before the beginning of the measurement.

The final diameter in case of measurement of samples from manufacturer C was set at 7.8 mm, because the radial force of the stents was so high the machine could not compress stents to 4 mm diameter (the resulting force was above the upper limit of the machine).

Table 2. Parameters of radial force measurement.

Setting Parameters	Manufacturer A	Manufacturer B	Manufacturer C	Manufacturer D
Initial diameter (mm)	25	25	25	28
Final diameter (mm)	4	4	7.8	4
Number of cycles	1	1	1	1

This research was conducted on 3 stent samples from each manufacturer and each sample was manufactured from different wire. Since the stent manufacturers did not provide a larger number of samples, it was not possible to replicate the measurements several times. Since the stents were manufactured using a standard manufacturing process, it can be assumed that this process is fully validated. A validated process gives repeatable results, so it can be concluded that the samples would also show repeatable and reproducible behavior.

3. Results

3.1. Surface State

The composition of the individual raw wires was 50.3 ± 0.6 at% Ti and 49.7 ± 0.6 at% Ni. Figure 1 shows the surface of all studied samples before the immersion in SGF using optical microscopy. The depth of the blue discoloration depends on the thickness of the TiO_2 layer caused by the heat treatment [33]. Factors for this are the concentration of oxygen during the annealing, the reactivity of the alloy, and any subsequent surface finishing. The darker color (dark gray and black) of the samples reflects the surface characteristics of the raw nitinol, as each stent manufacturer produced samples by their standard production process. Slight differences in the alloy result in different surface qualities, resulting in a darker color, which reflects the dark, black, or thin oxide layer [34–37]. The metallic color of some samples is caused by the surface treatment associated with the removal of oxidized material (e.g., grinding, polishing, electropolishing, chemical etching) [18].

Manufacturers A and B did not apply any surface finish and the different coloration is a consequence of different oxidization of the nitinol wire.

With manufacturer C, the surface is metallically shiny, regardless of the type of input wire, reflecting a more aggressive surface treatment in the finishing process, removing most of the passive layer.

Samples from manufacturer D show a greater variation in color, with the sample from wire 3 appearing metallically shiny, but the other two samples show different degrees of surface (dis-)coloration. These samples also underwent some surface treatment after annealing, but the quality of the raw material here is seen to have a greater effect on the final product.

The surface state was also studied by SEM (Figure 2). The raw wire 1 has an etched surface; on the raw wires 2 and 3, a thick oxide layer is visible. Based on the structure and color of the surface, the layer on wire 2 is thicker.

The influence of the processing is well visible. Manufacturers A and B use the wire in its original state. No other surface treatment except shape setting was used and the differences in surface structure are preserved. In the case of manufacturer C, the three samples look very similar with very little oxide layer on the surface, indicating the etching of the final stents. Manufacturer D also used etching but probably in weaker etchant because the original surface structure can still be identified.

Table 3 summarizes the results from XPS analysis. The chemical composition of the surface layer and the thickness of the TiO_2 layer were investigated simultaneously.

The original surfaces of the samples from stent manufacturers A, B. and D contained different concentrations of nickel depending on the type of wire used. In contrast, the chemical composition of the surface layer samples from manufacturer C only showed minimal differences. These samples, together with sample D3, showed the thinnest TiO_2 layer. This reflects the fact that these manufacturers apply an additional surface treatment in the finishing process of the stent. Whether other manufacturers use any surface treatment in the production process cannot be determined on the basis of this analysis and this information was not provided by them.

The surface concentration of nickel decreases with the application of chemical etching [38]. After immersion in SGF, the concentration of nickel decreases, indicating that it is preferentially released from the surface of the material.

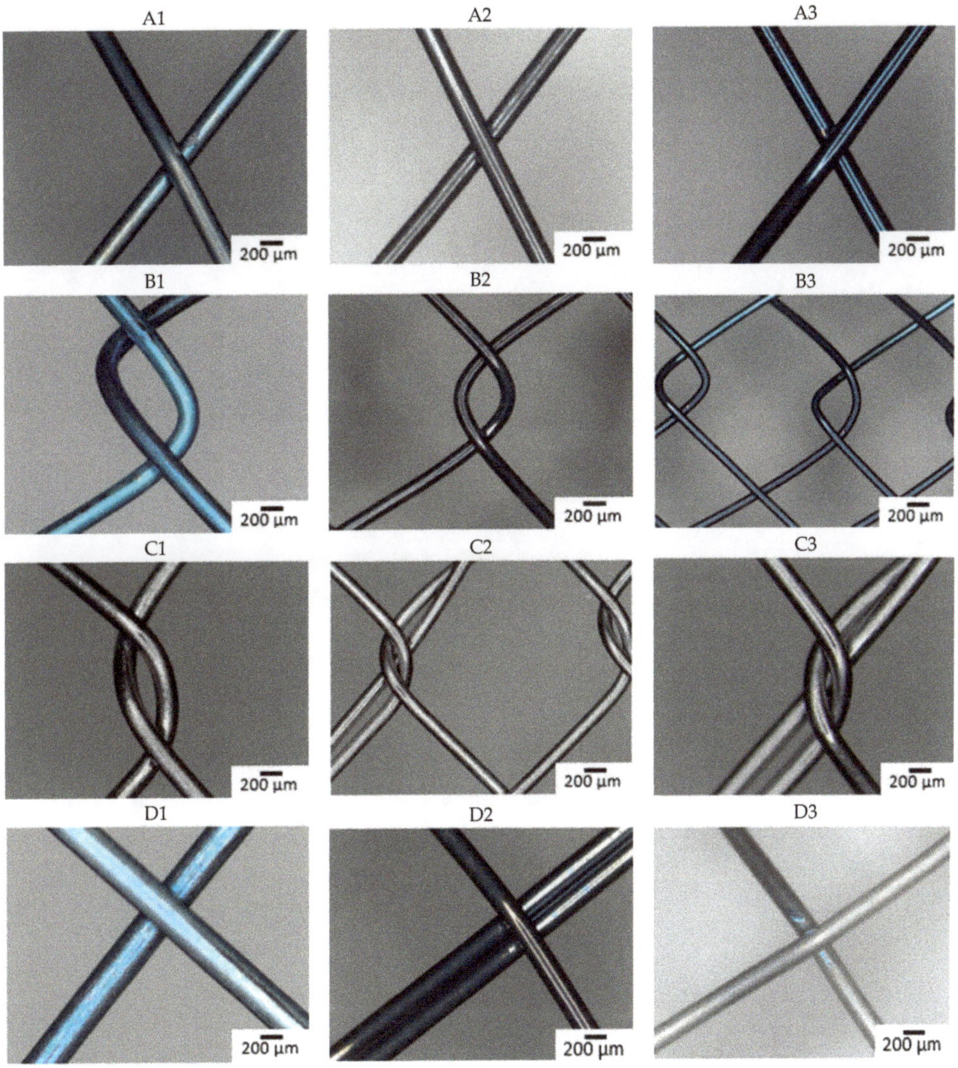

Figure 1. Optical microscopy images of the stents surface before immersion (**A–D**: individual manufacturers, **1–3**: different raw wire).

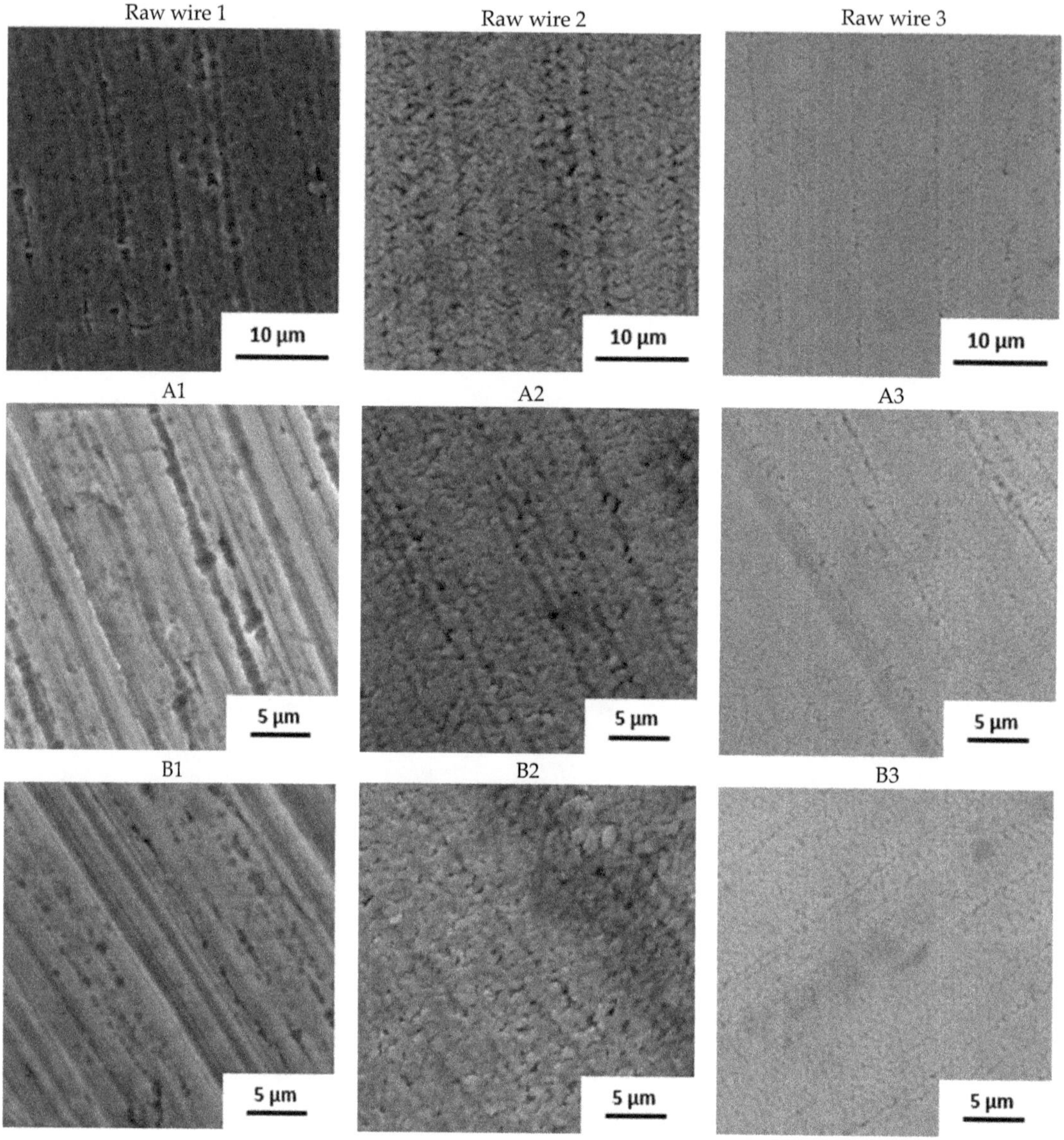

Figure 2. *Cont.*

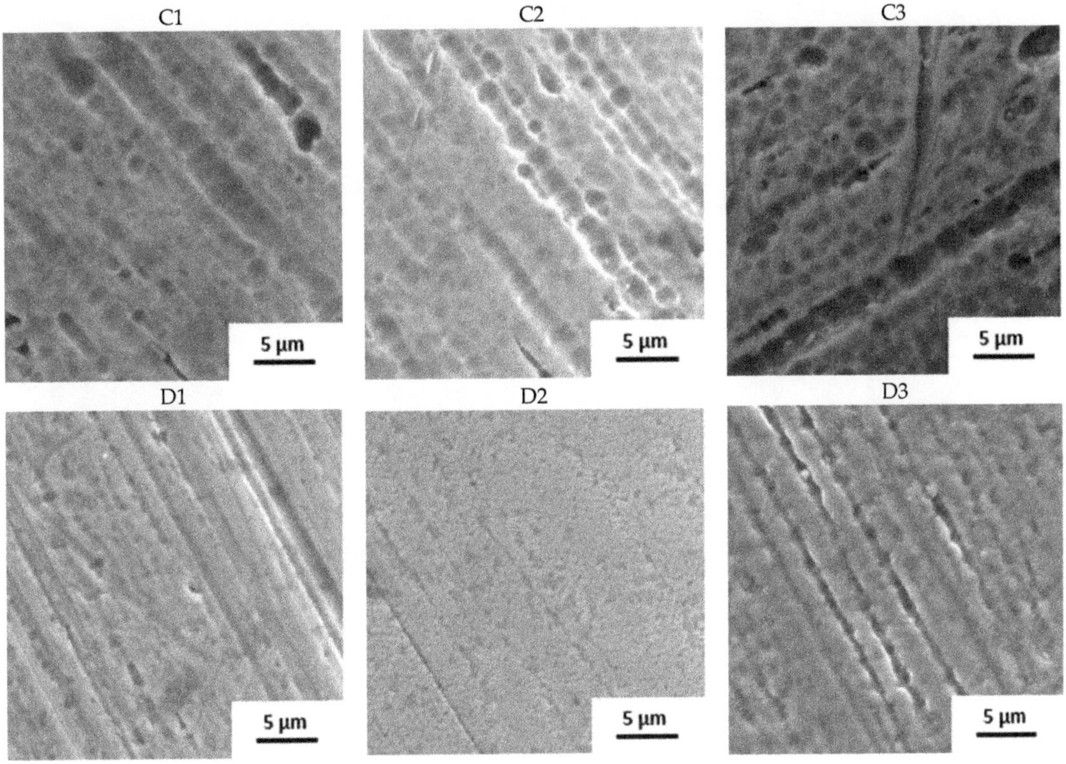

Figure 2. Surface of samples before the immersion in SGF studied by SEM (**A–D**: individual manufacturers, **1–3**: different raw wire).

Table 3. XPS analysis before and after 6-week immersion in SGF.

Sample	Original Surface			After 6 Weeks		
	Ti (at. %)	Ni (at. %)	TiO_2 Thickness (nm)	Ti (at. %)	Ni (at. %)	TiO_2 Thickness (nm)
wire 1	96.8	3.2	8.8	-	-	-
wire 2	67.4	32.6	≥10	-	-	-
wire 3	98.3	1.7	≥10	-	-	-
A1	96.6	3.4	≥10	97	3	≥10
A2	85.8	14.2	≥10	100	0	≥10
A3	81.4	18.6	≥10	100	0	≥10
B1	94.5	5.5	≥10	95.1	4.9	≥10
B2	82.1	17.9	≥10	89.3	10.7	9
B3	88.5	11.5	≥10	94.8	5.2	≥10
C1	92.2	7.8	8.7	94	6	8.9
C2	92.5	7.5	7.8	93.1	6.9	8.2
C3	94.4	5.6	7.8	92.5	7.5	7.8
D1	89.7	10.3	≥10	95.3	4.7	≥10
D2	94.8	5.2	≥10	97	3	≥10
D3	86.6	13.4	6.9	91.3	8.7	8.4

3.2. Immersion in SGF—Corrosion Behavior

Figure 3 shows the surface of all studied samples after the 6-week immersion in SGF using optical microscopy. Degradation due to corrosion of the material was most extensive on a sample from manufacturer B made of wire 2. A corrosion attack was initiated on a "free" surface, not at the intersection of individual wires. Other samples showed no corrosion pits or pits were so small that they were undetectable by optical microscopy. To better understand the effect of the combination of input material quality and manufacturing process on the corrosion resistance of the input material, the samples were subjected to further analyses (XPS).

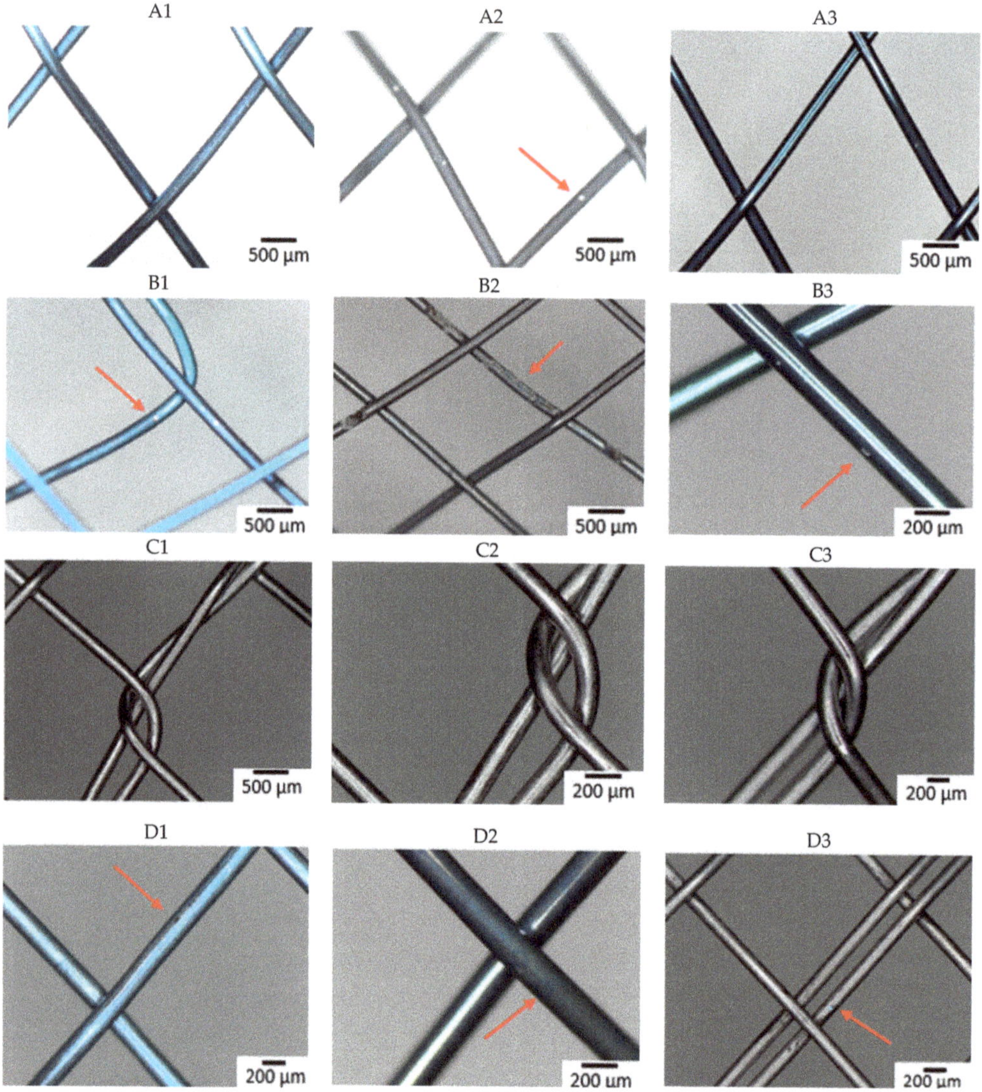

Figure 3. Surface of samples after 6-week immersion in SGF studied by optical microscopy. The arrow indicates the corrosion damage on the wire (**A–D**: individual manufacturers, **1–3**: different raw wire).

The time when the first pit was caused by the corrosion attack was also observed (see Table 4). All samples from manufacturer A showed corrosion attack after one week of exposure. This probably means that similar corrosion resistance can be expected regardless of the quality of the input material. In the case of samples from manufacturer B, the first pit appeared after two weeks in the case of wires 1 and 3. In the sample from wire 2, the first corrosion attack occurred after 3 days (first control of the samples). This means that from a corrosion resistance point of view, the production process depends on the quality of the input material. Samples from manufacturer C showed the longest delay to corrosion attack (wires 2 and 3) or not at all within 6 weeks (wire 1). This is likely due to the fact that the manufacturer applies a surface treatment during the production process, which thoroughly removes the oxide layer caused by the heat treatment. The layer formed during heat treatment is removed because it is not tight and blocks the possibility of creating a tight passive layer.

Table 4. Time to detection of first corrosion pit.

Manufacturer Wire	A	B	C	D
1	1 week	2 weeks	not detected	1 week
2	1 week	3 days	5 weeks	3 weeks
3	1 week	2 weeks	4 weeks	1 week

In the case of manufacturer D, wire 2 showed the longest resistance to the corrosion attack. This is in contrast with other manufacturers, where wire 2 showed poor corrosion resistance. This variation demonstrates the impact of the finishing process.

Based on the comparison with the E-pH diagram of NiTi at low pH, TiO_2 is formed and nickel in the form of Ni^{2+} ion dissolves [39]. The higher corrosion activity of NiTi alloy at lower pH was experimentally verified [40]. It should be noted that the E-pH diagram neglects the effect of the passive layer, which serves as protection against nickel dissolution. This is the main reason for detecting the first pit on samples at different times.

XPS analysis was also performed after immersion to monitor changes in surface chemical composition. All samples (except sample C3) had a decrease in Ni concentration on the surface (Table 3). Sample B2 showed the highest nickel concentration after exposure, which was probably due to the increased corrosion activity of this sample. Furthermore, there was no reduction in TiO_2 thickness of all samples after immersion.

AAS analysis of SGF was performed to determine the amount of Ni and Ti released from the materials after immersion (Table 5). Most nickel was released from the samples with the largest amount of that element in the surface layer before immersion (samples A2 and B2). Ti in SGF was also detected in these samples due to more extensive corrosion attack. The lowest amount of Ni after immersion was released from samples C1 to C3 (concentration was under detection limit of the machine). Vojtěch et al. reported that nickel ions release decreases with application of additional surface treatment after the heat treatment [38]. The range of the corrosion attack and first pitting detection was directly proportional to the amount of the released ions. It is evident especially for the Ni concentration. Ti was detected only in the case of an exerted corrosion attack.

Based on the surface images, evaluation of first corrosion pit and XPS analysis, it can be said that the Ni content on the surface has the greatest effect on the corrosion resistance of the material. We assume that the release or elution of Ni into the environment creates weaknesses in the passive layer leading to pit formation. This theory is supported by the fact that the attack did not occur at the point of wire crossing, but on the "free" surface, and therefore depending only on the quality of the surface layer, its chemical composition, and any inclusions or defects.

Table 5. AAS analysis of SGF after 6-week immersion.

Sample	Ni (ppm)	Ti (ppm)
A1	0.35	---
A2	3.93	2
A3	0.4	---
B1	0.18	---
B2	14.4	9
B3	0.37	---
C1	<0.1	---
C2	<0.1	---
C3	<0.1	---
D1	0.3	---
D2	0.5	---
D3	0.1	---

The corrosion resistance parameters are summarized in Table 6. The results show that the processing of the input material affected the corrosion resistance of the final product. While the wires of suppliers 1 and 2 showed an order-of-magnitude increase in corrosion rate due to processing, wire from supplier 2 showed a decrease (except for case B2). For wires 1 and 3, post-processing played a less significant role. The values of the polarization resistances for all manufacturers are approximately the same. For wire 2, processing no longer plays a significant role. For manufacturers C and D, the final products were probably etched (see Figure 2), which increased the corrosion resistance. Also, within individual wires, there are minimal differences between these manufacturers. For manufacturers A and B, the corrosion resistance is strongly dependent on the condition of the input material. The large variances in the individual measurements indicate uneven corrosion attack.

Table 6. Open circuit potential (OCP) and polarization resistance (Rp) after 1 h immersion in SGF. Data are presented as average ± confidence (α = 0.05).

OCP	Rp	OCP	Rp	OCP	Rp
mV/SSCE	Ω cm^2	mV/SSCE	Ω cm^2	mV/SSCE	Ω cm^2
A1		A2		A3	
−89 ± 9	$3{\cdot}10^5 \pm 8{\cdot}10^4$	−115 ± 45	$2{\cdot}10^4 \pm 1{\cdot}10^4$	−88 ± 12	$3{\cdot}10^5 \pm 2{\cdot}10^5$
B1		B2		B3	
37 ± 92	$5{\cdot}10^5 \pm 4{\cdot}10^5$	−63 ± 14	$4{\cdot}10^3 \pm 9{\cdot}10^2$	−71 ± 6	$9{\cdot}10^4 \pm 3{\cdot}10^4$
C1		C2		C3	
−55 ± 27	$6{\cdot}10^5 \pm 2{\cdot}10^5$	−60 ± 15	$6{\cdot}10^5 \pm 2{\cdot}10^5$	−42 ± 17	$5{\cdot}10^5 \pm 1{\cdot}10^5$
D1		D2		D3	
86 ± 21	$4{\cdot}10^5 \pm 1{\cdot}10^5$	35 ± 36	$4{\cdot}10^5 \pm 9{\cdot}10^4$	28 ± 10	$7{\cdot}10^5 \pm 2{\cdot}10^5$
1		2		3	
12 ± 61	$2{\cdot}10^6 \pm 1{\cdot}10^6$	−44 ± 45	$1{\cdot}10^4 \pm 6{\cdot}10^3$	91 ± 26	$2{\cdot}10^6 \pm 5{\cdot}10^5$

3.3. Immersion in SGF—Radial Forces Measurement

During the exposure, the samples were subjected to radial force measurements at regular intervals. The results of this measurement are summarized in Figure 4. It can be seen from the graphs that the radial resistive force (RRF) of all samples decreased during exposure. In the case of chronic outward forces (COF), the trend is no longer so clear-cut. For samples from manufacturer A, a slight decrease in this force was observed during the immersion. The same applies to samples from manufacturer B. However, the sample made of wire 2 showed a sharp decrease in COF after 4 weeks of immersion. Due to the low number of samples, it is not possible to state with sufficient accuracy whether this decrease was caused by a measurement error or a corrosion process.

In the case of samples from manufacturer C, COF stabilized after 3 days of immersion. This statement correlates with the fact that no signs of extensive corrosion were found on the surface of the samples after exposure.

The COF of the samples from manufacturer D did not show a significant change. The variance of values is probably due to the design of the samples. The samples did not have a tubular shape but were provided with flares which had a larger diameter than the stent body. These flares exerted a different force on the jaws of the measuring machine than the stent body, so the COF values showed greater variance.

The significant change of mechanical properties of sample B2 could be caused by exerted corrosion activity associated with hydrogen embrittlement, when potential may change locally, but not sufficiently enough to reflect on a macroscopic scale. Hydrogen embrittlement may occur upon immersion in an acidic environment [41]. Stent failure due to hydrogen embrittlement is also described by Volenec et al. [42].

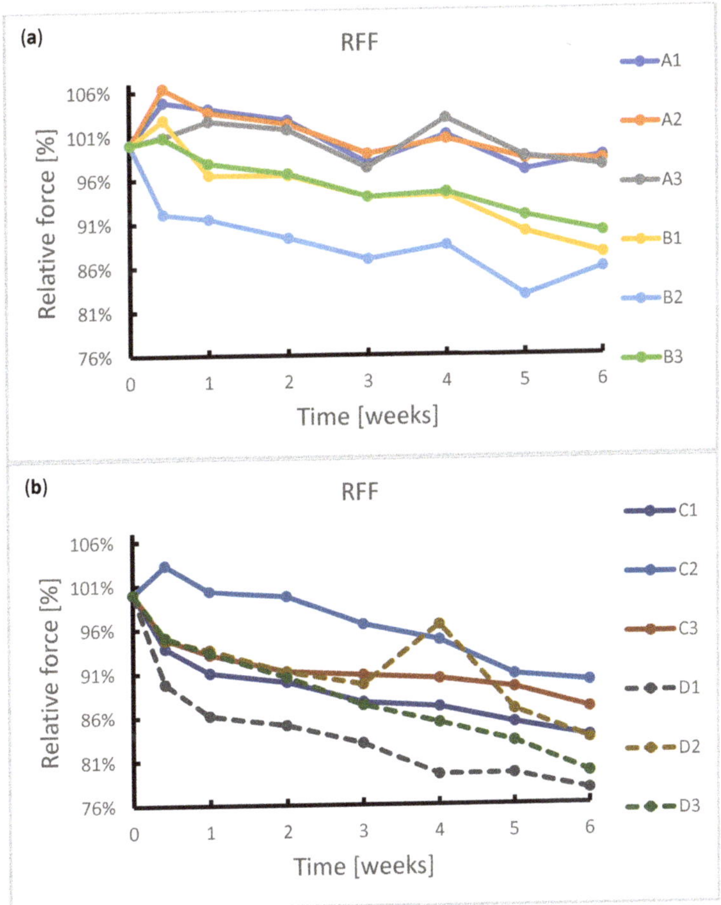

Figure 4. *Cont.*

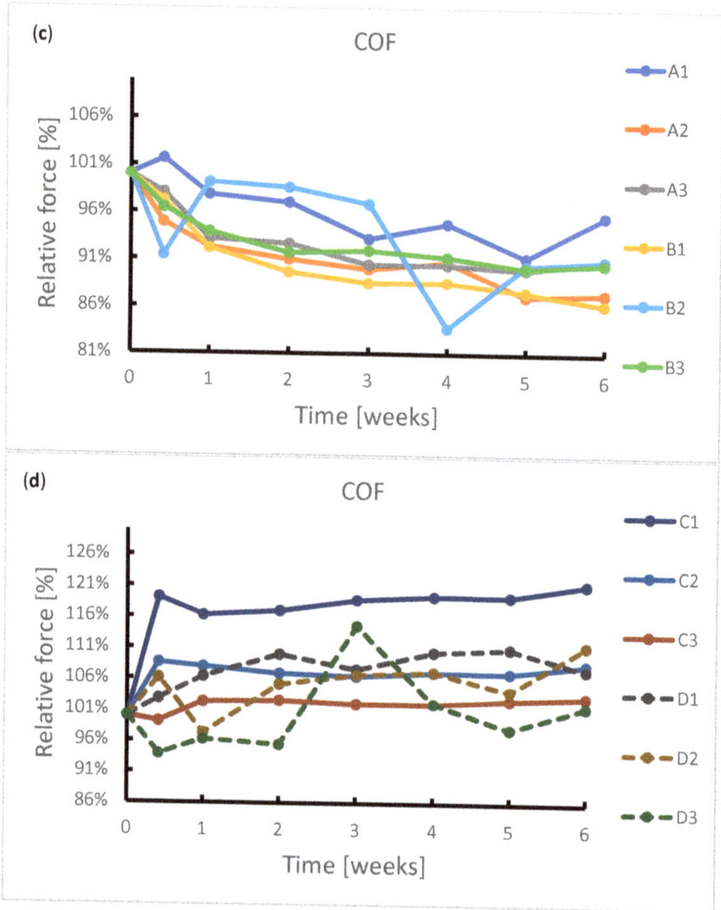

Figure 4. Reduction in radial forces (RRF—radial resistive force and COF—chronic outward force) during the 6-week immersion in SGF: (**a**) RRF of samples A1–B3, (**b**) RRF of samples C1–D3, (**c**) COF of samples A1–B3, (**d**) COF of samples C1–D3.

4. Discussion

The effect of the surface finish of stents on their corrosion susceptibility is poorly understood. Not all stent manufacturers submit the stents to corrosion testing and no specific guidelines exist. Where corrosion testing is undertaken it usually follows the ASTM standard F2129-19 ("Standard Test Method for Conducting Cyclic Potentiodynamic Polarization Measurements to Determine the Corrosion Susceptibility of Small Implant Devices"). However, this does not reflect the challenges presented to stents placed in or around the stomach, where they are exposed to a very low pH. For that reason, the ASTM standard F2528-06 ("Standard Test Methods for Enteral Feeding Devices with a Retention Balloon") was chosen, which was originally designed to assess the resilience of gastrostomy retention balloons. We recommend that future test standards take into account the additional challenges of a low pH environment, which accelerate corrosion. Unexpectedly, it was found that the corrosion attack is initiated on the "free" surface and not at the points of crossing of individual wires, where there are potential crevices. A higher amount of nickel in the surface layer has a negative effect on the corrosion resistance. In keeping with this, the range of corrosion attack and first pit detection was proportional to the amount of nickel ions released. Ti was detected only in the case of an exerted corrosion attack. This was also confirmed by the polarization resistance measurement, when the smallest polarization resistance was shown by the stent from which the most nickel was released into the environment. The best in vitro results were achieved with the combination of wire 1 and a surface finish of the final stent removing as much of the surface layer after heat treatment as possible. The results of that research provide initial results in the field of corrosion resistance of esophageal nitinol stents. It must be noted that in real life, additional factors play a role. Implanted stents are subjected to cyclic stress and exposure to a wide range of food and fluids, as well as bacterial flora. Furthermore, esophageal stents are very often provided with a covering which, among other things, reduces the exposure of the metal skeleton to gastric acid to a variable extent. For these reasons, stents may exhibit different corrosion resistance in vivo.

5. Conclusions

Our results demonstrate the effects of a combination of different input material treatment and manufacturing technology on the corrosion resistance of esophageal nitinol stents.

Significant differences in corrosion susceptibility were demonstrated between the three nitinol wires. In addition, after 6 weeks of immersion in SGF, it was found that production technology can increase or reduce the corrosion resistance further. Finishing techniques are able to reduce the impact of the quality of input materials on corrosion resistance, but the exact finishing methods of each manufacturer were not disclosed.

It was also found that an inappropriate combination of input wire and production process can lead to a significant deterioration in corrosion resistance. A stent surface that contains a larger amount of nickel on the surface after production process is much more susceptible to corrosion attack in simulated gastric fluid.

Etching appears to help improve corrosion resistance, but due to the nature of this study, it is not possible to suggest a precise treatment procedure for the stents.

Author Contributions: D.A.: Investigation, formal analysis, writing—original draft, writing—review and editing, funding acquisition; J.F.: Conceptualization, investigation, formal analysis, writing—original draft, writing—review and editing, visualization, supervision, project administration, funding acquisition; E.K.: Investigation, formal analysis; D.W.E.: Conceptualization, writing—original draft, validation, resources; H.-U.L.: Conceptualization, writing—original draft, validation, resources. All authors have read and agreed to the published version of the manuscript.

Funding: This research was funded by the Technology Agency of the Czech Republic, grant number FW03010583.

Data Availability Statement: The datasets used and analyzed during the current study are available from the corresponding author on reasonable request.

Acknowledgments: The authors would like to thank Ella-CS, MI Tech, S&G Biotech, and TaeWoong for their generous support.

Conflicts of Interest: The authors declare no conflict of interest.

References

1. Provan, J.L. Use of Celestin tube for palliation of malignant oesophageal obstruction. *Thorax* **1969**, *24*, 599–602. [CrossRef]
2. Hirdes, M.M.; Vleggaar, F.P.; de Beule, M.; Siersema, P.D. In vitro evaluation of the radial and axial force of self-expanding esophageal stents. *Endoscopy* **2013**, *45*, 997–1005. [CrossRef]
3. Mbah, N.; Philips, P.; Voor, M.J.; Martin, R.C.G., 2nd. Optimal radial force and size for palliation in gastroesophageal adenocarcinoma: A comparative analysis of current stent technology. *Surg. Endosc.* **2017**, *31*, 5076–5082. [CrossRef]
4. Laasch, H.-U.; Milward, G.D.; Edwards, D.W. 'Radial force' of colonic stents: A parameter without consistency, definition or standard. *Int. J. Gastrointest. Interv.* **2020**, *9*, 99–105. [CrossRef]
5. Hindy, P.; Hong, J.; Lam-Tsai, Y.; Gress, F. A comprehensive review of esophageal stents. *Gastroenterol. Hepatol.* **2012**, *8*, 526–534.
6. Liang, D.H.; Hwang, E.; Meisenbach, L.M.; Kim, M.P.; Chan, E.Y.; Khaitan, P.G. Clinical outcomes following self-expanding metal stent placement for esophageal salvage. *J. Thorac. Cardiovasc. Surg.* **2017**, *154*, 1145–1150. [CrossRef]

7. Stoeckel, D.; Pelton, A.; Duerig, T. Self-expanding nitinol stents: Material and design considerations. *Eur. Radiol.* **2004**, *14*, 292–301. [CrossRef]
8. Kocich, R.; Szurman, I.; Kursa, M. The methods of preparation of Ti–Ni–X alloys and their forming. In *Shape Memory Alloys-Processing, Characterization and Applications*; IntechOpen: London, UK, 2013; pp. 28–35. [CrossRef]
9. Shiva, S.; Palani, I.A.; Mishra, S.K.; Paul, C.P. Investigations on the influence of composition in the development of Ni–Ti shape memory alloy using laser based additive manufacturing. *Opt. Laser Technol.* **2015**, *69*, 44–51. [CrossRef]
10. Sun, F.; Jordan, L.; Albin, V.; Lair, V.; Ringuedé, A.; Prima, F. On the High Sensitivity of Corrosion Resistance of NiTi Stents with Respect to Inclusions: An Experimental Evidence. *ACS Omega* **2020**, *5*, 3073–3079. [CrossRef]
11. Kanca, M.S.; Kök, M.; Qader, I.N. Corrosion behavior and crystal-microstructural analysis of non-equiatomic NiTi shape memory alloys. *J. Therm. Anal. Calorim.* **2022**, *147*, 1–11. [CrossRef]
12. Martinez, A.L.; Saugo, M.; Flamini, D.O.; Saidman, S.B. Enhancing the corrosion behavior of Ti–6Al–4V and Nitinol alloys by simple chemical oxidation in H_2O_2. *Mater. Chem. Phys.* **2023**, *295*, 127069. [CrossRef]
13. Vojtech, D.; Joska, L.; Leitner, J. Influence of a controlled oxidation at moderate temperatures on the surface chemistry of nitinol wire. *Appl. Surf. Sci.* **2008**, *254*, 5664–5669. [CrossRef]
14. Vojtěch, D.; Voděrová, M.; Kubásek, J.; Novák, P.; Šedá, P.; Michalcová, A.; Fojt, J.; Hanuš, J.; Mestek, O. Effects of short-time heat treatment and subsequent chemical surface treatment on the mechanical properties, low-cycle fatigue behavior and corrosion resistance of a Ni–Ti (50.9at.% Ni) biomedical alloy wire used for the manufacture of stents. *Mater. Sci. Eng. A* **2011**, *528*, 1864–1876. [CrossRef]
15. Zhu, L.; Pelton, A.R.; Fino, J.M. Oxidation of nitinol and its effect on corrosion resistance. In Proceedings of the ASM Materials & Processes for Medical Device Conference, St. Paul, MN, USA, 25–27 August 2004; pp. 156–161.
16. Ohtsu, N.; Hirano, Y.; Yamaguchi, K.; Yamasaki, K. Surface characteristics, Ni ion release, and antibacterial efficacy of anodized NiTi alloy using HNO_3 electrolyte of various concentrations. *Appl. Surf. Sci.* **2019**, *492*, 785–791. [CrossRef]
17. Wu, G.; Li, L.; Sun, M.; Wang, Y.; Luo, F.; Zhang, Q.; Liu, R.; Chen, Z.; Yao, J. Microstructural evolution and biological properties of PEO coating on SLM-prepared NiTi alloy. *Surf. Coat. Technol.* **2023**, *452*, 129065. [CrossRef]
18. Hassel, A.W. Surface treatment of NiTi for medical applications. *Minim. Invasive Ther. Allied Technol.* **2004**, *13*, 240–247. [CrossRef]
19. Dos Reis Barros, C.D.; Gomes, J.A.d.C.P. Strain induced localized corrosion of NiTi, NiTiCo and NiTiCr alloys in 0.9% NaCl. *J. Mech. Behav. Biomed. Mater.* **2020**, *112*, 104015. [CrossRef]
20. Khara, H.S.; Diehl, D.L.; Gross, S.A. Esophageal stent fracture: Case report and review of the literature. *World J. Gastroenterol.* **2014**, *20*, 2715–2720. [CrossRef]
21. Wiedmann, M.; Heller, F.; Zeitz, M.; Mössner, J. Fracture of a covered self-expanding antireflux stent in two patients with distal esophageal carcinoma. *Endoscopy* **2009**, *41* (Suppl. S2), E129–E130. [CrossRef]
22. Rana, S.S.; Bhasin, D.K.; Sidhu, G.S.; Rawal, P.; Nagi, B.; Singh, K. Esophageal nitinol stent dysfunction because of fracture and collapse. *Endoscopy* **2009**, *41* (Suppl. S2), E170–E171. [CrossRef]
23. Wadsworth, C.A.; East, J.E.; Hoare, J.M. Early covered-stent fracture after placement for a benign esophageal stricture. *Gastrointest. Endosc.* **2010**, *72*, 1260–1261. [CrossRef] [PubMed]
24. Shokri, N.; Safavi, M.S.; Etminanfar, M.; Walsh, F.C.; Khalil-Allafi, J. Enhanced corrosion protection of NiTi orthopedic implants by highly crystalline hydroxyapatite deposited by spin coating: The importance of pre-treatment. *Mater. Chem. Phys.* **2021**, *259*, 124041. [CrossRef]
25. Nagaraja, S.; Sullivan, S.J.L.; Stafford, P.R.; Lucas, A.D.; Malkin, E. Impact of nitinol stent surface processing on in-vivo nickel release and biological response. *Acta Biomater.* **2018**, *72*, 424–433. [CrossRef] [PubMed]
26. Kay, L. Cyclic Potentiodynamic Survey of Medical Materials. Master's Thesis, Purdue University Fort Wayne Campus, Fort Wayne, IN, USA, 2010.
27. Carlson, T.A.; McGuire, G. Study of the X-ray photoelectron spectrum of tungsten—Tungsten oxide as a function of thickness of the surface oxide layer. *J. Electron Spectrosc. Relat. Phenom.* **1972**, *1*, 161–168. [CrossRef]
28. Strohmeier, B.R. An ESCA method for determining the oxide thickness on aluminum alloys. *Surf. Interface Anal.* **1990**, *15*, 51–56. [CrossRef]
29. *ASTM Standard F2528-06*; Standard Test Methods for Enteral Feeding Devices with a Retention Balloon. ASTM International: West Conshohocken, PA, USA, 2006.
30. Pandolfino, J.E.; Richter, J.E.; Ours, T.; Guardino, J.M.; Chapman, J.; Kahrilas, P.J. Ambulatory esophageal pH monitoring using a wireless system. *Am. J. Gastroenterol.* **2003**, *98*, 740–749. [CrossRef] [PubMed]
31. Washington, N.; Steele, R.J.; Wright, J.W.; Bush, D.; McIntosh, S.L.; Wilkinson, S.; Washington, C. An investigation of lower oesophageal redox potentials in gastro-oesophageal reflux patients and healthy volunteers. *Physiol. Meas.* **1997**, *18*, 363–371. [CrossRef]
32. Kaltsidis, H.; Mansoor, W.; Park, J.H.; Song, H.Y.; Edwards, D.W.; Laasch, H.U. Oesophageal stenting: Status quo and future challenges. *Br. J. Radiol.* **2018**, *91*, 20170935. [CrossRef]
33. De Almeida, B.; Elias, C. Influence of heat treatment on color and flexibility of nickel-titanium endodontic instruments. *RGO Rev. Gaúcha Odontol.* **2020**, *68*, e20200044. [CrossRef]
34. Furukawa Techno Material Co., Ltd. Ni-Ti Tubes for Medical Devices. Available online: https://www.furukawa-ftm.com/tokusyu/english/wp-content/themes/furukawa/assets/pdf/Ni-Ti_tube.pdf (accessed on 15 May 2023).

35. Fort Wayne Metals. Nitinol. Available online: https://www.fwmetals.com/materials/nitinol/ (accessed on 15 May 2023).
36. Patel, M.M.; Gordon, R.F. An investigation of diverse surface finishes on fatigue properties of superelastic Nitinol wire. In Proceedings of the International Conference on Shape Memory and Superelastic Technologies, SMST-2006, Pacific Grove, CA, USA, 7–11 May 2006, pp. 7–11.
37. PEIERTECH. Nitinol Round Wires Lingual Arch Wire NiTi Wire for Orthodontic. Available online: https://www.peiertech.cn/Nitinol-Round-Wires-Lingual-Arch-Wire-NiTi-Wire-for-Orthodontic-pd48474233.html (accessed on 15 May 2023).
38. Vojtěch, D.; Fojt, J.; Joska, L.; Novák, P. Surface treatment of NiTi shape memory alloy and its influence on corrosion behavior. *Surf. Coat. Technol.* **2010**, *204*, 3895–3901. [CrossRef]
39. Ding, R.; Shang, J.-X.; Wang, F.-H.; Chen, Y. Electrochemical Pourbaix diagrams of NiTi alloys from first-principles calculations and experimental aqueous states. *Comput. Mater. Sci.* **2018**, *143*, 431–438. [CrossRef]
40. Clarke, B.; Carroll, W.; Rochev, Y.; Hynes, M.; Bradley, D.; Plumley, D. Influence of Nitinol wire surface treatment on oxide thickness and composition and its subsequent effect on corrosion resistance and nickel ion release. *J. Biomed. Mater. Res. Part A* **2006**, *79A*, 61–70. [CrossRef] [PubMed]
41. Ogawa, T.; Yokozawa, E.; Oda, T.; Maruoka, K.; Sakai, J.i. Hydrogen embrittlement behavior of Ni-Ti shape memory alloy with different microstructures in acidic fluoride solution. *Int. J. Mech. Mater. Eng.* **2015**, *10*, 12. [CrossRef]
42. Volenec, K.; Pohl, I. The challenges: Stent materials from the perspective of the manufacturer. *Gastrointest. Interv.* **2016**, *5*, 98–104. [CrossRef]

metals

Article

Investigation on Blood Compatibility of Cu/Ti Metal Coating Prepared via Various Bias Voltages and Copper Content

Qiong Hu [1], Hengquan Liu [1,*], Fei Gao [2,*], Xi Yang [1], Junfeng Li [1], Ren Liu [3], Zexuan Liu [1] and Dongfang Wang [1]

[1] College of Materials and Chemistry & Chemical Engineering, Chengdu University of Technology, Chengdu 610059, China; qiongqhu@163.com (Q.H.); yangxi926494@163.com (X.Y.); lijunfeng@cdut.cn (J.L.); linzexuan@hotmai.com (Z.L.); wangdongfang2021@163.com (D.W.)
[2] Chengdu Neurotrans Medical Technology Co., Ltd., Chengdu 610094, China
[3] Department of Research & Development, Chengdu Medtech-Life Co., Ltd., Chengdu 610094, China; medtechlife@aliyun.com
* Correspondence: liuhengquan15@cdut.edu.cn (H.L.); phil.gao@neurotrans.com.cn (F.G.)

Abstract: Surface modification of some metal coatings is usually used to improve the blood compatibility of biomaterials; however, some aspects of the bological properties of metal coatings cannot be adjusted via the content of each component. In this work, Cu/Ti metal coatings with various amounts of copper content were prepared by the physical vapor deposition (PVD) method, and the influence of deposition bias was further investigated. Phase structure, element composition and surface morphology were investigated by X-ray diffraction (XRD), X-ray photoelectron spectroscopy (XPS) and scanning electron microscopy, respectively. The hemolysis ratio, platelet adhesion and protein adsorption were applied to evaluate the blood compatibility. The results show that a Cu/Ti coating of uniform quality can be obtained; the dispersion of the deposition and copper content is regulated by the number of copper sheets, but the deposition bias does not obviously affect the copper content of the Cu/Ti coating. The hemolysis rate of the Cu/Ti coating is less than 0.4%, the degree of platelet adhesion is significantly reduced on Cu/Ti coatings compared to control samples, and the contact angle of all coatings is greater than that of pure titanium. The largest adsorption capacity of BSA was found on the coating with the deposition bias voltage of −40 V. The number of copper flakes is increased, and the adsorption of FIB on the Cu/Ti coating surface is reduced. Therefore, Cu/Ti coatings prepared via this deposition method have potential for applications to regulate blood compatibility and surface performance.

Keywords: Cu/Ti coating; bias voltage; copper content; blood compatibility

Citation: Hu, Q.; Liu, H.; Gao, F.; Yang, X.; Li, J.; Liu, R.; Liu, Z.; Wang, D. Investigation on Blood Compatibility of Cu/Ti Metal Coating Prepared via Various Bias Voltages and Copper Content. *Metals* **2022**, *12*, 435. https://doi.org/10.3390/met12030435

Academic Editor: Sergey N. Grigoriev

Received: 21 January 2022
Accepted: 20 February 2022
Published: 1 March 2022

1. Introduction

Cardiovascular disease (CVD) is one of the major causes of human mortality around the world. The latest report shows that the death count caused by CVD is approximately half of all deaths [1]. The current, most effective treatment method of CVD is interventional therapy with a vessel stent. After the stent is implanted in the body, the surface of the stent material will interact with the physiological components; protein adsorption and activation, blood cell destruction and an intimal increase occur at the interface between the material and blood. In order to improve the blood compatibility of the stent surface, some researchers have reported that surface modification of inorganic coatings of the stent surface can be employed to improve blood compatibility, such as a Ti-O coating [2–4], TiN coating [5,6], diamond-like carbon [7–12], etc. Although the blood compatibility has been improved via these inorganic coatings, the above coatings cannot be widely used in vascular stents because of their brittleness.

Titanium alloys have been widely used as biological materials on account of their low density, non-magnetic nature [13], light weight [14], good corrosion resistance [15], high fatigue strength [16] and good biocompatibility [17,18]. Copper is an essential element for the normal growth of organisms and the development of tissue structure [19–21], and it plays an important role in the homeostasis, inflammation, proliferation and remodeling of wound healing. H Liu et al. [22,23] successfully prepared Cu/Ti coatings by vacuum arc source deposition and physical vapor deposition. The study found that Cu/Ti coatings were less toxic to endothelial cells and showed good biocompatibility. At the same time, Cu/Ti coatings present a good inhibitory effect, but the Cu/Ti coatings prepared by vacuum arc plasma deposition easily contain micro-sputtered particles, which affect the continuity of the damaged coatings. D Wojcieszak et al. [24] reported that Cu/Ti coatings were prepared by magnetron co-sputtering (Ti target and Cu target); the result showed that Cu/Ti coatings were less toxic towards fibroblasts. While their work focused on the bactericidal activity and cytotoxicity of the Cu/Ti coating, its blood compatibility was not mentioned, and there are higher requirements for the magnetron sputtering equipment.

In order to obtain a uniform and smooth Cu/Ti metal coating, the magnetron sputtering of a single target was employed to prepare it, and the copper content and morphology of the Cu/Ti composite coating were regulated by changing the number of copper sheets and the bias voltage. The blood compatibility of Cu/Ti coatings with different deposition bias and copper content was also investigated.

2. Materials and Methods

2.1. Preparation of Cu/Ti Metal Coatings

Single target sputtering by magnetron sputtering was adopted to prepare Cu/Ti composite coatings. A certain number and size of copper sheets were placed on the titanium target. The copper–titanium composite target with different copper–titanium content could be prepared by adjusting the quantity and size of copper sheets. Cu/Ti composite coatings with different copper content were prepared under a certain sputtering process.

The Cu/Ti metal coating was prepared on a medical titanium surface (10 mm × 10 mm) by magnetron sputtering. A titanium target with a diameter of 50.0 mm and a purity of 99.99% and a copper sheet with a diameter of 10 mm and a purity of 99.999% were used in the preparation process. Prior to the deposition of the coating, the surface of the medical titanium substrate was polished with different waterproof sandpapers; then, the substrate was ultrasonically cleaned in acetone and anhydrous ethanol for 10 min, and the treated medical titanium was placed in a vacuum chamber. At the beginning of coating preparation, the pressure in the vacuum chamber was pumped to 6×10^{-4} Pa by a mechanical pump and molecular pump, and the flow rate of argon gas was 120 SCCM. When the pressure of the vacuum chamber reached 2.0 Pa, a bias of −700 V was applied, and the surface of the sample was bombarded with argon ions for 10 min. Regarding the deposition process, the technical parameters of sputtering and detailed data are shown in Table 1.

Table 1. The preparation parameters of various Cu/Ti coatings.

Sample No	Sputtering Pressure (Pa)	Sputtering Power (W)	Sputtering Bias (V)	Deposition Time (min)	Copper Number (Slice)
1#	0.4	100	0	60	1
2#	0.4	100	−40	60	1
3#	0.4	100	−80	60	1
4#	0.4	100	−120	60	1
5#	0.4	100	−40	60	2
6#	0.4	100	−40	60	3

2.2. Coating Characterization

The surface morphology and phase structure of the coatings was characterized by field emission scanning electron microscopy (SEM, FEI INSPECT F50, Hillsboro, OR, USA) and X-ray diffraction (XRD, D2 PHASER, Bruker, Billerica, MA, USA). X-ray photoelectron spectroscopy (XPS, VG Science, East Grinstead, UK) was used to detect the chemical composition and chemical valence state of the coating surface and the interior. To study the wettability of the coating surface, a liquid–solid wetting angle analyzer (SDC-200, SINDIN, Dongguan, China) was used to examine the hydrophilicity and hydrophobicity of the coating.

2.3. Blood Compatibility

2.3.1. Hemolysis Ratio

The hemolysis assay was performed in accordance with ISO 109934:2009 (National standard for hemolysis rate). We added 8 mL of anticoagulant blood to a container containing 10 mL 0.9% sodium chloride injection, and gently shook the mixture to prepare diluted blood; the absorbance at 545 nm was obtained with an ultraviolet spectrophotometer. The hemolysis ratio was calculated according to the following formula:

$$HR\ (\%) = (A - A_1)/(A_2 - A_1) \times 100\% \tag{1}$$

where HR is the hemolysis ratio (%), A is the absorbance of the sample (%), A_1 is the absorbance of the negative controls (%) and A_2 is the absorbance of the positive control (%).

2.3.2. Platelet Adhesion

We centrifuged the fresh human blood and removed the supernatant. The coated sample was ultrasonically cleaned with absolute ethanol and distilled water. The sample was soaked in platelet-rich plasma (PRP), and then incubated in a constant-temperature water bath at 37 °C for 2 h. After rinsing the samples with phosphate-buffered saline (PBS), the samples were fixed with 2% glutaraldehyde for 4 h, and then dehydrated with ethanol at concentrations of 50%, 70%, 90% and 100%. After drying, the platelet adhesion number and morphology were observed under the scanning electron microscope.

2.3.3. Platelet Activation

PRP was obtained by centrifugation of fresh whole blood containing sodium citrate. We spread the platelet-rich plasma fully on the surface of the sample (10 mm × 10 mm), incubated it at 37 °C for 2 h and washed it with bovine serum albumin solution 3 times. Subsequently, mouse anti-human CD62P antibody (primary antibody) was added, the sample was incubated at 37 °C for 1 h, and then a secondary antibody solution was added (goat anti-mouse polyclonal antibody labeled with horseradish peroxidase) and incubated at 37 °C for 1 h. Finally, we added the chromogenic agent and the material for 5–10 min, transferred the chromogenic solution to the microtiter plate, added sulfuric acid solution to stop the color, placed the sample in the microplate reader and calculated the platelet activation amount according to the optical density (OD) value.

2.4. Protein Adsorption

The Protein Quantification Kit (BCA Assay) was used to measure the adsorption behavior of the two main proteins (albumin, fibrinogen and globulin) in the plasma on the surface of the material. First, we prepared 1 mg/mL protein solution with PBS solution, equilibrated the coating surface (10 mm × 10 mm) with PBS solution for 2 h and then immersed it in 3 mL protein (albumin, fibrinogen, globulin) solution and allowed it to adsorb at 37 °C for 2 h. The coating was immersed in 2% sodium dodecyl sulfonate (SDS) solution and stirred at 37 °C for 2 h to fully resolve the surface-absorbed proteins into the SDS solution. The number of absorbed proteins was determined by using the corresponding kit for reaction and coloration.

3. Results and Discussion

3.1. Structure and Surface Topography of Thin Films

The XRD patterns of various Cu/Ti composite coatings are shown in Figure 1. The position of the titanium peak (100) of samples 2# and 3# is shifted from Figure 1a, and the diffraction peaks of Ti are broadened more obviously. The broadening of the diffraction peaks may be because Cu is solid dissolved in Ti crystals to form a solid solution of Ti, causing lattice distortion, peak broadening and peak shifting.

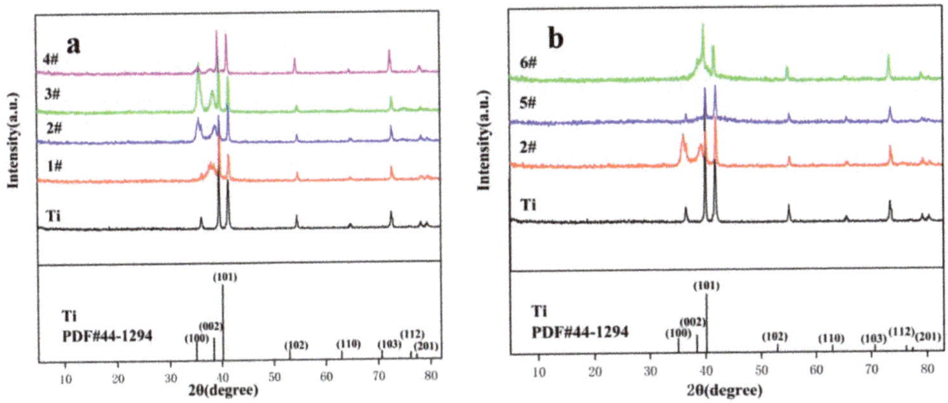

Figure 1. XRD pattern of Cu/Ti coating: (**a**) different bias voltage, (**b**) different number of copper sheets.

The number of copper sheets increases, the (100) peak is gradually reduced (Figure 1b), and the center of the XRD broadening peak gradually shifts to the right, which may be due to the formation of some Cu/Ti phases containing Cu at the same time [25], and it gradually increases with the increase in the copper content in the film.

Figure 2 shows the high-resolution spectra of Ti2p of samples 2# and 4#. The peaks of Ti2p1/2 and Ti2p3/2 appear at 464.1 eV and 458.6 eV, respectively, and the difference in the electron binding energy between the two is 5.7 eV, which indicates that the Ti on the film surface is oxidized into TiO_2 in the air. The Cu2p peak binding energies of elemental Cu and CuO only differ by 0.1 eV [26]. It is difficult to distinguish elemental Cu from Cu_2O by the Cu2p peak. However, it is reported in the literature that Cu on the surface of a film is easily oxidized, and Cu should exist in two forms, CuO and Cu_2O, and not contain elemental Cu [24].

Table 2 shows the element content of 1#, 2#, 3# and 4# samples. We found that with the increase in deposition bias, the content of copper and titanium did not change significantly.

Table 2. The element content of 1#, 2#, 3# and 4# samples.

Sample	Element Component (at%)		
	Ti	Cu	O
1#	20.04	0.49	79.47
2#	21.34	0.52	78.14
3#	22.91	0.54	76.55
4#	23.77	0.53	75.77

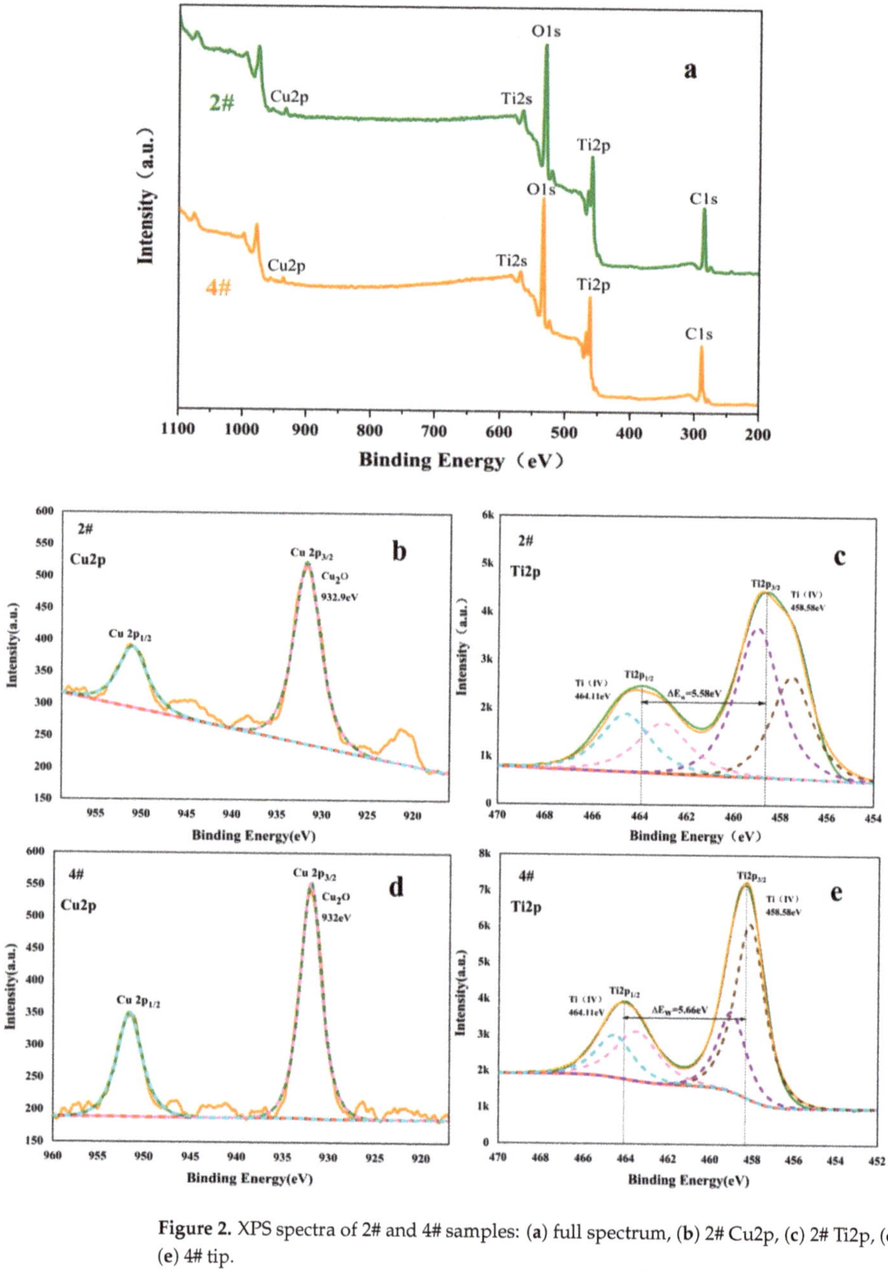

Figure 2. XPS spectra of 2# and 4# samples: (**a**) full spectrum, (**b**) 2# Cu2p, (**c**) 2# Ti2p, (**d**) 4# Cu2p, (**e**) 4# tip.

The SEM morphology of the Cu/Ti composite coatings is shown in Figure 3. It can be clearly observed that the surface of the Cu/Ti composite coating is uniform, and grain boundaries and obvious cracks are not found. All coatings show directional streaks, which may be caused by grinding of the substrate. The surface smoothness of the Cu/Ti coating first increases and then decreases following the increase in the bias voltage. A suitable energy is conducive to the migration of sputtered atoms under a suitable deposition bias;

thereby, the surface smoothness would be increased. Some particles were observed on the surface of the 4# coating and distinct crystal boundaries between the particles were also found. These results might be due to the introduction of bias, leading to the enhancement of the energy deposition effect, which was conducive to the growth of surface grains. It can be seen from Figure 3e–g that the density of sample 2# is higher than that of samples 5# and 6#: this is because the sputtering energy of copper is relatively low, so the Cu content of sample 6# is higher than that of sample 2# and sample 5#.

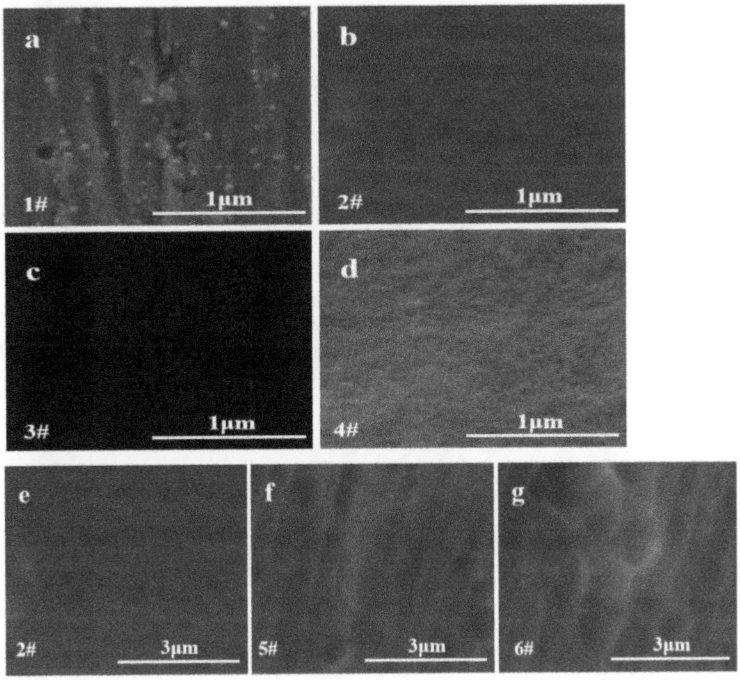

Figure 3. The SEM images of various Cu/Ti composite coatings: (**a**) 1#, (**b**) 2#, (**c**) 3#, (**d**) 4#, (**e**) 2#, (**f**) 5#, (**g**) 6#.

3.2. Surface Wetting Test

The biological behaviors of cell adhesion, proliferation and migration on the material surface are affected by the wettability of the material surface [26,27]. The surface wettability angle and surface energy of various coatings are shown in Figure 4a,b, respectively. The contact angle of medical titanium is the smallest, which is approximately 67°. The contact angle of the Cu/Ti composite coating is larger than that of medical titanium.

The free energy of the Cu/Ti composite coating is less than that of medical titanium, as shown in Figure 4c,d. The higher the surface free energy of the coating, the smaller the contact angle and the better the water droplet wettability.

The hemolysis rates of all Cu/Ti coatings was less than 0.4, which is in accordance with the ISO 10993-4:2002 standard; it indicates that all Cu/Ti composite coatings were safe.

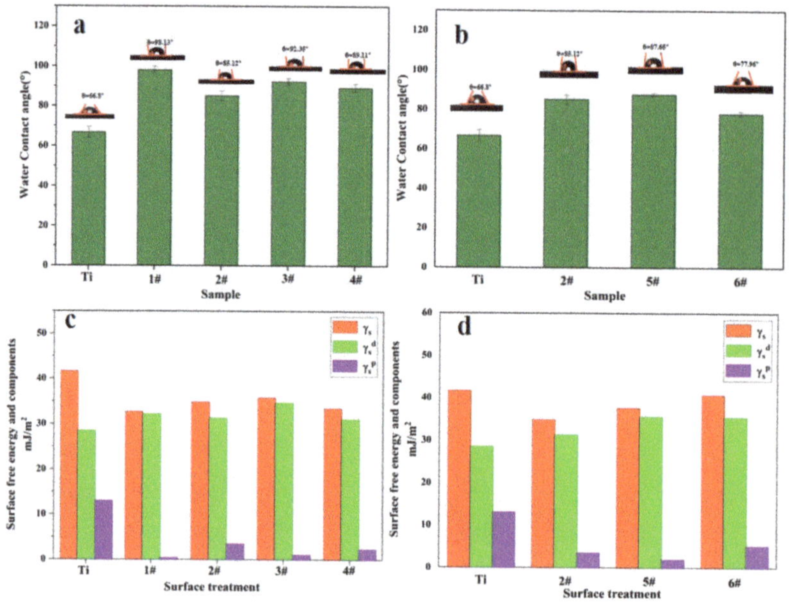

Figure 4. The contact angle and surface energy of various samples: (**a**) the contact angle of 1#–4#, (**b**) the contact angle of 2#, 5# and 6#, (**c**) free energy of 1#–4#, (**d**) free energy of 2#, 5# and 6#.

3.3. Platelet Adhesion and Activation

Platelet adhesion and activation are the main causes of thrombosis. The platelet adhesion on the surface of medical titanium is shown in Figure 5a. Platelets spread across a large area were found, and the platelet aggregation and activation were very evident. Figure 5b–e show that, compared to medical titanium, the adhesion and aggregation of platelets on the Cu/Ti coating surface prepared under a certain deposition bias were reduced. It can be seen from Figure 5g,h that the number of platelets adsorbed by the samples is significantly reduced, and the shape is almost round, without aggregation and spreading behavior.

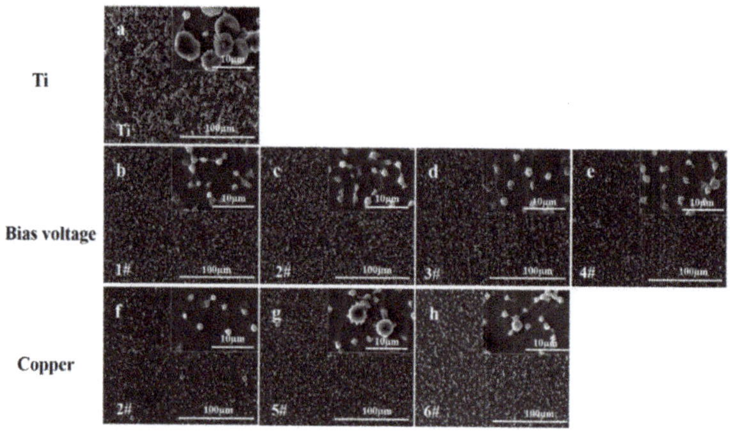

Figure 5. Platelet adhesion of Cu/Ti composite coatings: (**a**) Ti, (**b**) 1#, (**c**) 2#, (**d**) 3#, (**e**) 4#, (**f**) 2#, (**g**) 5#, (**h**) 6#.

The activation of platelets of various samples is described in Figure 6. These samples were different from the Cu/Ti coatings prepared with various parameters such as bias voltage and number of copper sheets. It can be seen from Figure 6a that the deposition bias is increased, and the effect of inhibiting platelet activation becomes greater, which may be related to the density and smoothness of the coating surface. Figure 6b shows the platelet activation under different numbers of copper plates. The platelet activation on the surfaces of samples 2#, 5# and 6# is inhibited. As the number of copper plates increases, the amount of platelet activation shows an increasing trend.

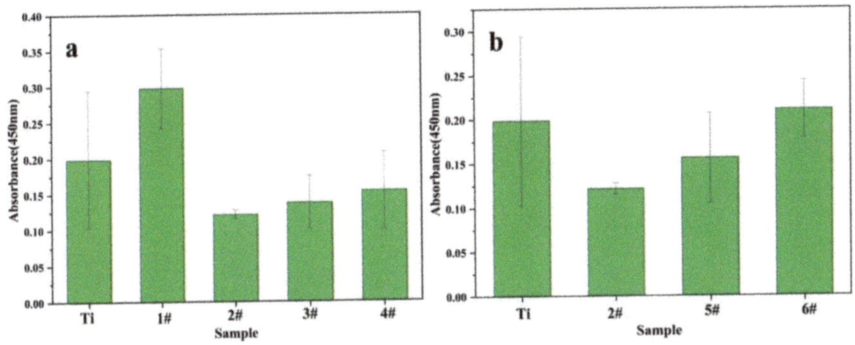

Figure 6. The platelet activation. (**a**) Different bias voltage, (**b**) different copper content.

3.4. Protein Adsorption

The Protein Quantification Kit (BCA Assay) method was used to further study the adsorption of proteins. The adsorption of albumin can reduce the adhesion and activation of platelets; thereby, the coagulation performance of the material's surface would be inhibited [28]. On the contrary, the blood compatibility of the material would be reduced owing to the adsorption and denaturation of fibrinogen (FIB) [29]. The adsorption of Bulked Segregant Analysis (BSA) and FIB on the Cu/Ti composite coating is shown in Figure 7. It can be seen from Figure 7a that the amount of BSA adsorbed on the surface of sample 4# is the largest, and the amount of FIB adsorbed on the surface of sample 2# is the largest. It can be found that the adsorption amount of BSA on the surfaces of samples 2#, 5# and 6# increases with the increase in copper content, as shown in Figure 7b.

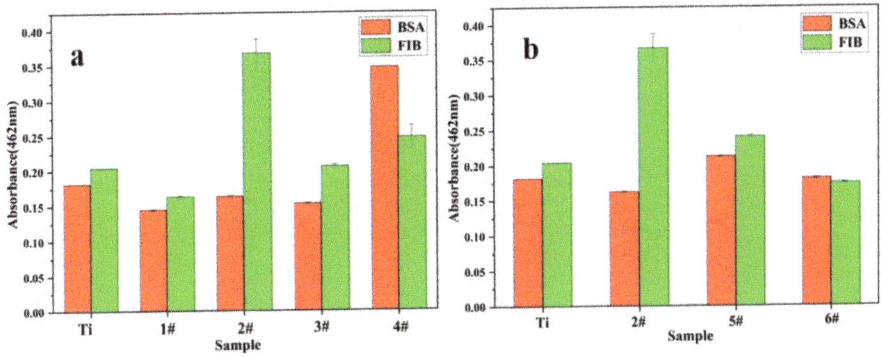

Figure 7. Surface fibrinogen and albumin adsorption capacity on various surfaces (indicated by OD value). (**a**) Different bias voltage, (**b**) different copper content.

4. Conclusions

In this study, a Cu/Ti composite coating was successfully prepared by magnetron single-target sputtering, and its surface wettability, protein adsorption and blood compatibility were affected by the deposition bias and copper content. The main conclusions are as follows.

(1) The copper content of various coatings is mainly determined by the number of copper sheets; the deposition bias was detected as a due to the influence of the quality of the coating.
(2) As the deposition bias increases, the smoothness and density of the coating are first increased and then decreased.
(3) The maximum amount of BSA adsorption occurs when the deposition bias voltage is −40 V. With the increase in the number of copper sheets, the adsorption amount of FIB gradually decreases.
(4) The Cu/Ti coatings prepared with different bias voltages and different numbers of copper sheets significantly reduced platelet adhesion, and the degree of platelet activation on the coating surface gradually increased.
(5) The results demonstrate that Cu/Ti composite coatings could improve the blood compatibility; this research result can be used for the surface modification of blood contact materials.

Author Contributions: Conceptualization, H.L. and F.G.; methodology, Q.H.; software, Q.H.; validation, X.Y., D.W. and Z.L.; formal analysis, Q.H.; investigation, J.L.; resources, H.L., R.L. and F.G.; data curation, Q.H.; writing—original draft preparation, Q.H.; writing—review and editing, H.L. and Q.H.; visualization, H.L.; supervision, H.L. and J.L.; project administration, H.L.; funding acquisition, H.L. All authors have read and agreed to the published version of the manuscript.

Funding: This work was jointly and financially supported by the Sichuan Science and Technology Program (Grant No. 2020JDRC0070). This study was also supported by the Jiangsu Province Engineering Research Center for Biomedical Materials and Advanced Medical Devices.

Data Availability Statement: Not applicated.

Conflicts of Interest: The authors declare no conflict of interest.

References

1. Arora, S.; Stouffer, G.A.; Kucharska-Newton, A.M.; Qamar, A.; Vaduganathan, M.; Pandey, A.; Porterfield, D.; Blankstein, R.; Rosamond, W.D.; Bhatt, D.L.; et al. Twentyyear trends and sex differences in young adults hospitalized with acute myocardial infarction. *Circulation* **2019**, *139*, 1047–1056. [CrossRef] [PubMed]
2. Zong, M.; Bai, L.; Liu, Y.; Wang, X.; Zhang, X.; Huang, X.; Hang, R.; Tang, B. Antibacterial ability and angiogenic activity of Cu–Ti–O nanotube arrays. *Mater. Sci. Eng. C* **2017**, *71*, 93–99. [CrossRef] [PubMed]
3. Liu, Y.; Hang, R.; Zhao, Y.; Bai, L.; Sun, Y.; Yao, X.; Jia, H.; Tang, B.; Hang, R. The effects of annealing temperature on corrosion behavior, Ni2+ release, cytocompatibility, and antibacterial ability of Ni–Ti–O nanopores on NiTi alloy. *Surf. Coat. Technol.* **2018**, *352*, 175–181. [CrossRef]
4. Hang, R.; Liu, Y.; Bai, L.; Zong, M.; Wang, X.; Zhang, X.; Huang, X.; Tang, B. Electrochemical synthesis, corrosion behavior and cytocompatibility of Ni–Ti–O nanopores on NiTi alloy. *Mater. Lett.* **2017**, *202*, 5–8. [CrossRef]
5. Lin, N.; Huang, X.; Zou, J.; Zhang, X.; Qin, L.; Fan, A.; Tang, B. Effects of plasma nitriding and multiple arc ion plating TiN coating on bacterial adhesion of commercial pure titanium via in vitro investigations. *Surf. Coat. Technol.* **2012**, *209*, 212–215. [CrossRef]
6. Hussein, M.A.; Ankah, N.K.; Kumar, A.M.; Azeem, M.A.; Saravanan, S.; Sorour, A.A.; Al Aqeeli, N. Mechanical, biocorrosion, and antibacterial properties of nanocrystalline TiN coating for orthopedicapplications. *J. Ceram. Int.* **2020**, *46*, 18573–18583. [CrossRef]
7. Liao, T.T.; Zhang, T.F.; Li, S.S.; Deng, Q.Y.; Wu, B.J.; Zhang, Y.Z.; Zhou, Y.J.; Guo, Y.B.; Leng, Y.X.; Huang, N. Biological responses of diamond-like carbon (DLC) films with different structures in biomedical application. *J. Mater. Sci. Eng. C* **2016**, *69*, 751–759. [CrossRef]
8. Choudhury, D.; Ching, H.A.; Mamat, A.B.; Cizek, J.; Abu Osman, N.A.; Vrbka, M.; Hartl, M.; Krupka, I. Fabrication and characterization of DLC coated microdimples on hip prosthesis heads. *J. Biomed. Mater. Res. Part B* **2015**, *103*, 1002–1012. [CrossRef]
9. Zhang, T.F.; Deng, Q.Y.; Liu, B.; Wu, B.J.; Jing, F.J.; Leng, Y.X.; Huang, N. Wear and Corrosion Properties of diamond like carbon (DLC) Coating on stainless steel, CoCrMo and Ti6Al4V substrate. *Surf. Coat. Technol.* **2015**, *273*, 12–19. [CrossRef]

10. Dhandapani, V.S.; Subbiah, R.; Thangavel, E.; Arumugam, M.; Park, K.; Gasem, Z.M.; Veeraragavan, V.; Kim, D.-E. Tribological properties, corrosion resistance and biocompatibility of magnetron sputtered titanium-amorphous carbon coatings. *Appl. Surf. Sci.* **2016**, *371*, 262–274. [CrossRef]
11. Bociaga, D.; Sobczyk-Guzenda, A.; Szymanski, W.; Jedrzejczak, A.; Jastrzebska, A.; Olejnik, A.; Swiatek, L.; Jastrzebski, K. Diamond like carbon coatings doped by Si fabricated by a multi-target DC-RF magnetron sputtering method—Mechanical properties, chemical analysis and biological evaluation. *Vacuum* **2017**, *143*, 395–406. [CrossRef]
12. Peng, F.; Lin, Y.; Zhang, D.; Ruan, Q.; Tang, K.; Li, M.; Liu, X.; Chu, P.K.; Zhang, Y. Corrosion Behavior and Biocompatibility of Diamond-like Carbon-Coated Zinc: An In Vitro Study. *J. ACS Omega* **2021**, *6*, 9843–9851. [CrossRef] [PubMed]
13. Sha, X.; Xiao, N.; Guan, Y.; Yi, X. A first-principles investigation on mechanical and metallic properties of titanium carbides under pressure. *Mater. Sci. Technol.* **2018**, *34*, 1953–1958. [CrossRef]
14. Xiangyu, Z.; Meng, L.; Xiaojing, H.; Ruiqiang, H.; Xiaobo, H.; Yueyue, W.; Xiaohong, Y.; Bin, T. Antibacterial activity of single crystalline silver-doped anatase TiO_2 nanowire arrays. *J. Appl. Surf. Sci.* **2016**, *372*, 139–144.
15. Trino, L.D.; Dias, L.F.; Albano, L.G.; Bronze-Uhle, E.S.; Rangel, E.C.; Graeff, C.F.; Lisboa-Filho, P.N. Zinc Oxide Surface Functionalization and Related Effects on Corrosion Resistanceof Titanium Implants. *Ceram. Int.* **2018**, *44*, 4000–4008. [CrossRef]
16. Wang, R.; He, X.; Gao, Y.; Zhang, X.; Yao, X.; Tang, B. Antimicrobial property, cytocompatibility and corrosion resistance of Zn-doped ZrO_2/TiO_2 coatings on Ti6Al4V implants. *J. Mater. Sci. Eng. C* **2017**, *75*, 7–15. [CrossRef]
17. Chiang, H.-J.; Chou, H.-H.; Ou, K.-L.; Sugiatno, E.; Ruslin, M.; Waris, R.A.; Huang, C.-F.; Liu, C.-M.; Peng, P.-W. Evaluation of Surface Characteristics and Hemocompatibility on the Oxygen Plasma-Modified Biomedical Titanium. *Metals* **2018**, *8*, 513. [CrossRef]
18. Hu, H.; Zhang, W.; Qiao, Y.Q.; Jiang, X.Y.; Liu, X.; Ding, C. Antibacterial activity and increased bone marrow stem cell functions of Zn-incorporated TiO_2 coatings on titanium. *Acta Biomater.* **2012**, *8*, 904–915. [CrossRef]
19. Li, Y.; Liu, L.; Wan, P.; Zhai, Z.; Mao, Z.; Ouyang, Z.; Yu, D.; Sun, Q.; Tan, L.; Ren, L.; et al. Biodegradable Mg-Cu alloy implants with antibacterial activity for the treatment of osteomyelitis: In vitro and in vivo evaluations. *Biomaterials* **2016**, *106*, 250–263. [CrossRef]
20. Pohanka, M. Copper and copper nanoparticles toxicity and their impact on basic functions in the body. *Bratisl. Med. J.* **2019**, *120*, 397–409. [CrossRef]
21. Kornblatt, A.P.; Nicoletti, V.G.; Travaglia, A. The neglected role of copper ions in wound healing. *J. Inorg. Biochem.* **2016**, *161*, 1–8. [CrossRef] [PubMed]
22. Liu, H.; Zhang, D.; Shen, F.; Zhang, G.; Song, S. Hemocompatibility and anti-endothelialization of copper-titanium coating for vena cava filters. *J. Surf. Coat. Technol.* **2012**, *206*, 3501–3507. [CrossRef]
23. Liu, H.; Zhang, D.; Shen, F.; Zhang, G.; Song, S. Corrosion and ion release behavior of Cu/Ti film prepared via physical vapor deposition in vitro as potential biomaterials for cardiovascular devices. *J. Appl. Surf. Sci.* **2012**, *258*, 7286–7291. [CrossRef]
24. Wojcieszak, D.; Kaczmarek, D.; Antosiak, A.; Mazur, M.; Rybak, Z.; Rusak, A.; Osekowska, M.; Poniedzialek, A.; Gamian, A.; Szponar, B. Influence of Cu-Ti thin film surface properties on antimicrobial activity and viability of living cells. *J. Mater. Sci. Eng. C* **2015**, *56*, 48–56. [CrossRef]
25. Stranak, V.; Wulff, H.; Rebl, H.; Zietz, C.; Arndt, K.; Bogdanowicz, R.; Nebe, B.; Bader, R.; Podbielski, A.; Hubicka, Z.; et al. Deposition of thin titanium- copper films with antimicrobial effect by advanced magnetron sputtering methods. *Mater. Sci. Eng. C* **2011**, *31*, 1512–1519. [CrossRef]
26. Ghodselahi, T.; Vesaghi, M.A.; Shafiekhani, A.; Baghizadeh, A.; Lameii, M. XPS study of the Cu@Cu_2O core-shellnanoparticles. *J. Appl. Surf. Sci.* **2008**, *255*, 2730–2734. [CrossRef]
27. Wu, Y.; Simonovsky, F.I.; Ratner, B.D.; Horbett, T.A. The role of adsorbed fibrinogen in platelet adhesion to polyurethane surfaces: A comparison of surface hydrophobicity, protein adsorption, monoclonal antibody binding, and platelet adhesion. *J. Biomed. Mater. Res. A* **2005**, *74*, 722–738. [CrossRef]
28. Kim, C.H.; Khil, M.S.; Kim, H.Y.; Lee, H.U.; Jahng, K.Y. An improved hydrophilicity via electrospinning for enhanced cell attachment and proliferation. *J. Biomed. Mater. B* **2006**, *78*, 283–290. [CrossRef]
29. Liu, H.Q.; Siedlecki, C.A. (Eds.) *Hemocompatibility of Biomaterials for Clinical Applications*; Woodhead Publishing: Thorston, UK, 2018; pp. 379–394.

MDPI
St. Alban-Anlage 66
4052 Basel
Switzerland
www.mdpi.com

Metals Editorial Office
E-mail: metals@mdpi.com
www.mdpi.com/journal/metals

www.ingramcontent.com/pod-product-compliance
Lightning Source LLC
Chambersburg PA
CBHW062038220926
51850CB00010B/2599
* 9 7 8 3 7 2 5 8 0 0 3 2 2 *